Nursing School Entrance Exam Review Supplement

Prepared by

Jennifer P. Harris
Portland Community College

THOMSON

BROOKS/COLE

Australia • Canada • Mexico • Singapore • Spain • United Kingdom • United States

For more information about our products,
contact us at:
Thomson Learning Academic Resource Center
1-800-423-0563

For permission to use material from this text or
product, submit a request online at
http://www.thomsonrights.com.
Any additional questions about permissions can be
submitted by email to **thomsonrights@thomson.com.**

Thomson Brooks/Cole
10 Davis Drive
Belmont, CA 94002-3098
USA

Asia
Thomson Learning
5 Shenton Way #01-01
UIC Building
Singapore 068808

Australia/New Zealand
Thomson Learning
102 Dodds Street
Southbank, Victoria 3006
Australia

Canada
Nelson
1120 Birchmount Road
Toronto, Ontario M1K 5G4
Canada

Europe/Middle East/South Africa
Thomson Learning
High Holborn House
50/51 Bedford Row
London WC1R 4LR
United Kingdom

Latin America
Thomson Learning
Seneca, 53
Colonia Polanco
11560 Mexico D.F.
Mexico

Spain/Portugal
Paraninfo
Calle/Magallanes, 25
28015 Madrid, Spain

Table of Contents

In-Depth Solutions 71

Exam Preparation Tips

1. Study throughout the year, semester, and/or term.
    a. The longer your preparation time, the more relaxed you will feel preparing for the exam.
    b. Revisiting the information will allow you to incorporate this knowledge into your long term memory.
    c. Studying frequently for shorter time periods can be more effective than studying periodically for longer time periods.
    d. Make a study schedule. Follow the schedule for a week, reflect on how well the schedule is working, and make adjustments as needed.

2. Review the "Key Ideas" sections.
    a. The ideas are separated by a theme.
    b. They are also linked to specific sections from two different chemistry textbooks:
        i. <u>Chemistry for Today</u>, 5th ed., by Seager and Slabaugh
        ii. <u>Introduction to General, Organic, and Biochemistry</u>, 7th ed., by Bettelheim, Brown, and March

3. Keep a record of your study time.
    a. Working out your answers in a specific notebook will be helpful when you perform your final review.
    b. Writing a brief journal entry to yourself before and/or after your study time will also be helpful as you review key ideas for the exam.

4. Answer the practice questions, check your work, and look at the in-depth solutions.

5. Make notes about trouble spots and talk with your chemistry instructor or a tutor.

6. Practice similar problems in your textbook or another exam preparation guide.

7. Remember, this guide will help you prepare for the questions related to the chemistry portions of your exam. You will likely need another exam preparation guide to help with other portions of the exam.

Tips for the Week before the Exam

1. Look back over your "trouble spot" notes. Review the material.

2. Avoid cramming. Ideally, you have already prepared for several weeks and/or months. This is the time to brush up on a few details, not start your studying.

3. Be fully rested for your exam.

4. Eat a healthy breakfast before your exam.

Key Ideas	Topics for Review	*Chemistry For Today*[1]	*Introduction To GOB*[2]
Mathematics and Measurement	Measurement Units	1.6	1.4
	Metric Prefixes	1.6	1.4
	Area	1.6	1.4
	Volume	1.6	1.4
	Fractions	1.9, 1.10, 1.11	1.5, 1.7
	Percentages	1.10, 2.7	2.4
	Dimensional Analysis (Factor-Unit Method)	1.6, 1.9	1.5
	Temperature Scales	1.6	1.4
Matter	Classification	1.4	1.6, 2.2, 2.5
	Chemical/Physical Properties	1.2, 3.3, 3.5, 3.6	1.1,
	Elements	1.4, 2.1	2.5
	Compounds	1.4, 2.1	3.5, 3.7
	Mixtures	1.4	2.2
	Isotopes	2.3, 2.5	2.4
	Atoms	1.3, 2.2, 2.3	2.1, 2.4, 3.2
Periodic Table	Symbols	2.1, 2.3, 2.5, 2.7	2.4, 2.5
	Periods	3.1	2.5
	Families	3.1, 3.5	2.5
	Blocks/Areas	3.5	2.7
	Special Groups	3.5	2.5,
	Trends	3.6, 4.2, 4.9	2.5, 2.7, 2.8
	Valence	3.5, 3.6, 5.3	2.6
	Electron configurations	3.4	2.6
Compounds	Nomenclature	4.4, 4.10	3.3, 3.6, 3.8
	Masses/Weights	2.4, 2.5, 2.6, 2.7, 4.5	2.4, 4.2, 4.3,
	Ions	2.2, 3.6, 4.2, 4.3, 4.10	3.2
	Bonding	4.2, 4.3, 4.6, 4.7	3.1, 3.4, 3.5, 3.7
	Interparticle Forces	4.11	5.7, 6.6
	Structure	4.5, 4.6, 4.7, 4.8, 4.9	3.5, 3.7, 3.9, 3.10
Chemical Reactions	Classification Schemes	5.2, 5.3, 5.4, 5.5, 5.6	4.4, 4.7
	Energy	5.8	4.8
	Activity Series	9.6	8.6
	Solubility Rules	7.3	6.4
	Predicting Products	5.3, 5.4, 5.5, 5.6, 9.6, 9.7, 9.8	8.6
	Balancing	5.1	4.4
	Oxidation Numbers	5.3	4.7
	Stoichiometry	5.9, 5.10, 5.11, 7.6, 9.10, 9.11	4.5

[1] Chemistry for Today: General, Organic, and Biochemistry, 5th ed., ©2004. by Spencer L. Seager and Michael R. Slabaugh

[2] Introduction to General, Organic, and Biochemistry, 7th ed., ©2004. by Frederick A. Bettelheim, William H. Brown, and Jerry March.

Key Ideas	Topics for Review	*Chemistry For Today*	*Introduction To GOB*
States of Matter and Phase Changes	STP	6.8	5.4
	Molar Volume	6.8	5.4
	Pressure	6.6	5.2
	Partial Pressure	6.9	5.5
	Composition of Air	6.9	Problem 5.41
	Combined Gas Law	6.7	5.3
	Graham's Law	6.10	1.8
	Phase Changes	6.11, 6.12, 6.13, 6.14, 6.15	5.9, 5.11
	Specific Heat	6.15	1.9, 5.11
	Energy	6.2	1.8
	Kinetic Molecular Theory	6.2	5.1, 5.6
Solutions	Saturation	7.2	6.4
	Solubility Trends	7.2	6.4
	Solution Concentrations	7.4	6.5
	Electrolytes	7.7	6.6
	Colligative Properties	7.7	6.8
	Osmosis	7.7	6.8
Acids, Bases, and Salts	Acid Properties	9.6	8.6
	Base Properties	9.7	8.6
	Strong vs. Weak Acids	9.9	8.2
	Conjugate Acid-Base Pairs	9.2	8.3
	H^+/OH^- Concentrations	9.4	8.7
	pH scale	9.5	8.8
	Salts	9.8	8.9
	Indicators	9.10	8.8
	Metal/Nonmetal Oxides in Water	9.8	8.6
	Buffers	9.13	8.11
Nuclear Chemistry	Decay Processes	10.2	9.1, 9.2
	Properties of Radiation	10.1	9.1
	Decay Equations	10.2	9.2
	Half-life	10.3	9.3
	Fusion	10.9	9.7
	Fission	10.9	9.8
	$E=mc^2$	10.9	
Organic Chemistry	Organic vs. Inorganic	11.2	10.1
	Key Functional Groups	11.4	10.4
	General Formulas	11.5, 12.5, 13.7	11.2
	Bond Length vs. Strength	4.6	3.7
	Important Organic Compounds	12.8, 13.5, 14.4, 15.4, 16.5, 16.6	11.11, 12.5, 14.6,
	Properties of Organic Compounds	11.10, 12.3, 13.3, 13.6, 13.8, 14.2, 15.2 , 16.3, 16.8	11.8, 12.4, 14.2, 14.4, 16.4, 17.4, 18.2

Key Ideas	Topics for Review	*Chemistry For Today*	*Introduction To GOB*
Carbohydrates	Classification	17.4, 17.7, 17.8	19.2, 19.5, 19.6
	Reducing/Nonreducing sugars	17.7	19.4
	Important carbohydrates	17.6, 17.7, 17.8	19.5
Lipids	Classification	18.1	20.1
	Cell Membranes	18.8	20.5
	Cholesterol	18.9	20.9
	Steroid Hormones	18.10	20.10
	Bile	18.9	20.11
	Important Lipids	18.4-18.11	20.6-20.12
Proteins	Structure	19.6-19.9	21.7-21.9
	Function	19.5	21.1
	Classification	19.5	21.1
	Important Proteins	19.4, 19.5	21.6
	Amino Acids	19.1	21.2
	Enzymes	20.1	22.1
	Enzyme Activity	20.5	22.4
	Mechanism of Enzyme Action	20.4	22.5
	Optimum Enzyme Conditions	20.6	22.4
	Common Enzymes	20.9	22.7
Nucleic Acids	Structure	21.1, 21.2	24.2, 24.3
	Function	21.3	24.1
	Location	21.3	24.1
	Complimentary base pairs	21.2	24.3
	Genetic engineering	21.10	24.8
			Chemical Connections
	Human Genome Project	21.3	24 D
	Mutations	21.9	9.5
Nutrition and Energy	Caloric need	22.1	29.3
	Sources of vitamins and minerals	22.3, 22.4	29.7
	Sources of biomolecules	22.2	29.2
	Essential amino acids	22.2	29.6
	Biomolecules and Energy	22.6	27.3-27.8
	Metabolism	22.6, 23.1, 24.3, 24.8	26.1, 27.1
	Fat-soluble vitamins	22.3	29.7
	Water-soluble vitamins	22.3	29.7
	Elements in the body	22.4	29.7
	Mineral needs/usage	22.4	29.7
	Photosynthesis	22.5	28.2
	Cellular respiration	22.7	26.2
	Krebs Cycle	23.5	26.4
	Electron transport chain	23.6	26.5

Key Ideas	Topics for Review	*Chemistry For Today*	*Introduction To GOB*
Body Fluids	Composition	25.1	31.1
	Urine	24.5	31.5
	Blood	25.6-25.9	31.2
	CO_2/O_2 transport	25.2	31.3-31.4
	pH regulation	25.5-25.10	31.6
	Diabetes	Chemistry and Your Health: 23.1	Chemical Connections 27 C

Vocabulary List

absorb
acid
acidosis
activation energy
active transport
ADP
aerobic respiration
affinity
albuminuria
alcohol
aldehyde
alkaline
alkalosis
alkane
alkene
alkyne
alpha particle
amide
amine
amino acid
anaerobic respiration
anion
aromatic
atmosphere
atom
ATP
barometer
base
beta particle
biomolecule
boiling point
buffer
calorie
Calorie
carbohydrate
carboxylic acid
catalyst
cation
Celsius
Central Dogma of Molecular
Biology
cholesterol
combination

combustion
compound
concentrated
covalent bond
deamination
decomposition
density
dialysis
diffusion
digestion
dilute
direct relationship
disaccharide
dissociate
dissolve
double bond
double replacement
ductile
electron
electron capture
Electron Transport Chain
electronegativity
element
empirical formula
endergonic
energy level
entropy
enzyme
essential amino acid
ester
ether
evaporation
exergonic
facilitated transport
Fahrenheit
fats
fermentation
fission
flammable
formula weight
fusion
gamma ray
gluconeogenesis

glucosuria
glycogenolysis
glycogensis
glycolysis
gram molecular weight
half-life
hemoglobin
heterogeneous
homogeneous
hydrogen bonding
hydrolysis
hydrophilic
hydrophobic
hypertonic
hypotonic
immiscible
inert gas
initiate
inorganic
insoluble
insulin
inverse relationship
ion
ionic bond
ionization energy
isomer
isotonic
isotope
Kelvin
ketoacidosis
ketone
kinetic energy
Krebs cycle
lipid
litmus
London dispersion force
malleable
melting point
miscible
mixture
mmHg
molality
molarity

mole
molecular weight
molecule
molten
monosaccharide
mutation
neutron
noble gas
non-electrolyte
nonpolar
normality
nucleotide
organic
osmosis
osmotic pressure
oxidation
oxidative phosphorylation
peptide bond
percentage
pH
photosynthesis
plasma
polar
polymer
polymerase chain reaction

polypeptide
polysaccharide
potential energy
precipitation
property
protein
proton
radius
reactive
reduction
renal
replication
respiration
salt
satiated
saturated
semipermeable
single bond
single replacement
soap
soluble
solute
solution
solvent
specific heat

STP
strong acid
strong electrolyte
sugar
supersaturated
symmetry
terminate
theory
thermometer
torr
transcription
transition element
translation
triple bond
unsaturated
urea
valence
vapor pressure
weak acid
weak electrolyte
weigh

Mathematics and Measurement

Directions: Each question or incomplete statement below is followed by four suggested answers of completions. Select the best answer choice.

1. The area of the shaded portion of the rectangle below is

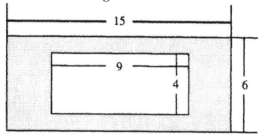

   (A) 54 square inches.
   (B) 90 square inches.
   (C) 45 square inches
   (D) 36 square inches.

2. The area of a mirror 40 inches long and 20 inches wide is approximately
   (A) 8.5 square feet.
   (B) 5.5 square feet.
   (C) 8.0 square feet.
   (D) 2.5 square feet.

3. A floor is 25 feet wide by 36 feet long. To cover this floor with carpet will require
   (A) 100 square yards.
   (B) 300 square yards.
   (C) 900 square yards.
   (D) 25 square yards.

4. What is the width of a rectangle with an area of 63 square feet and a length of 9 feet?
   (A) 22.5 feet
   (B) 7 feet
   (C) 567 feet
   (D) 144 feet

5. If the area of a square is 144 m², what is the length of a side of the square?
   (A) 12 m
   (B) 12 m²
   (C) 72 m
   (D) 72 m²

6. The number of cubic feet of soil needed for a flower box 3 feet long, 8 inches wide, and 1 foot deep is
   (A) 24
   (B) 12
   (C) $4\frac{2}{3}$
   (D) 2

7. What is the volume of a sphere of a radius 3 centimeters?
   (A) 119.05 cc
   (B) 113.04 cc
   (C) 106.00 cc
   (D) 101.08 cc

8. If a barrel has a capacity of 100 gallons, how many gallons will it contain when it is two-fifths full?
   (A) 20 gallons
   (B) 60 gallons
   (C) 40 gallons
   (D) 80 gallons

9. If an IV bag has a capacity of 1,260 milliliters, how many milliliters does it contain when it's two-thirds full?
   (A) 809 ml.
   (B) 750 ml.
   (C) 630 ml.
   (D) 840 ml.

10. The freshman nursing class consists of 40 students. $\frac{7}{8}$ of the class are women. How many women are in the class?
    (A) 35
    (B) 38
    (C) 21
    (D) 24

1

11. Mary drank 8 ounces of milk from a quart containing 32 ounces of milk. What part of the quart had she consumed?

   (A) $\frac{1}{4}$ (B) $\frac{1}{6}$ (C) $\frac{1}{3}$ (D) $\frac{1}{2}$

12. There are 75 nursing students in the freshman class, and 15 are men. What is the ratio of women to men?

   (A) $\frac{1}{4}$ (B) $\frac{1}{5}$ (C) $\frac{4}{1}$ (D) $\frac{5}{1}$

13. If a recipe calls for 5 ounces of sugar for every 15 ounces of flour, what part of the mixture will be sugar?

   (A) $\frac{1}{2}$ (B) $\frac{1}{5}$ (C) $\frac{1}{3}$ (D) $\frac{1}{4}$

14. How many grains of codeine are there in $1\frac{1}{2}$ tablets of $\frac{1}{8}$ grain each?

   (A) $\frac{2}{8}$ (B) $\frac{3}{16}$ (C) $\frac{3}{8}$ (D) $\frac{5}{8}$

15. The density of gold (Au) is 19.3g/cm³ and that of iron (Fe) is 7.9g/com³. A comparison of the volumes (V) of 50 gram samples of each metal would show that

   (A) $V_{Au} = V_{Fe}$ (C) $V_{Au} > V_{Fe}$
   (B) $V_{Au} < V_{Fe}$ (D) There is no predictable relationship between volumes.

16. If your monthly electricity bill increases from $80 to $90, the percentage of increase is, most nearly,

   (A) 10 percent. (B) $11\frac{1}{9}$ percent. (C) $12\frac{1}{2}$ percent. (D) $14\frac{1}{7}$ percent.

17. If a nursing exam contained 80 questions and you answered 72 of them correctly, what percent of the questions did you answer correctly?

   (A) 90 percent (B) 72 percent (C) 8 percent (D) 28 percent

18. If a monthly salary of $3,000 is subject to a 20 percent tax, the net salary is

   (A) $2,000 (B) $2,400 (C) $2,500 (D) $2,600

19. If a boat is purchased for $21,500 and sold for $23,650, what is the percentage of gain?

   (A) 8 percent (B) 15 percent (C) 20 percent (D) 10 percent

20. After deducting a discount of $16\frac{2}{3}$ percent, the price of a blouse was $35. The list price was

   (A) $37.50 (B) $38 (C) $41.75 (D) $42

21. Jane Doe borrowed $225,000 for five years at $13\frac{1}{2}$ percent. The annual interest charge was

   (A) $1,667 (B) $6,000 (C) $30,375 (D) $39,375

22. A junior salesman gets a commission of 14 percent on his sales. If he wants his commission to amount to $140, he will have to sell merchandise totaling

   (A) $1,960 (B) $10 (C) $1,000 (D) $100

23. A fashionable dress shop offers a 20 percent discount on selected items. For a dress marked at $280, what is the discount price?

   (A) $224.00 (B) $232.00 (C) $248.00 (D) $261.00

24. A store sold jackets for $65 each. The jackets cost the store $50 each. The percentage of increase of selling price over cost is
    (A) 40 percent.   (B) $33\frac{1}{2}$ percent.   (C) $33\frac{1}{3}$ percent.   (D) 30 percent.

25. If 24 percent of the students who enrolled in an algebra class of 50 students dropped the course before the semester ended, how many students remained in the class?
    (A) 27   (B) 76   (C) 12   (D) 38

26. Find .2 percent of 400.
    (A) 80   (B) 800   (C) .8   (D) 2000

27. An 8-ounce bottle of fruit juice provides 200 calories. What percent of the 200 calories is provided by 3 ounces of the fruit juice?
    (A) 12.5 percent   (B) 22.5 percent   (C) 37.5 percent   (D) 75 percent

28. Consider the circle with central angles shown below:

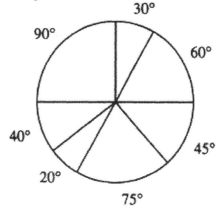

    What percent of the circle does the central angle of 60° represent? (Give answer to the nearest degree.)
    (A) 6 percent   (B) 60 percent   (C) 17 percent   (D) 25 percent

29. What percentage of 20 is 12?
    (A) 12 percent   (B) 60 percent   (C) 14 percent   (D) 8 percent

30. Three percent of the 900 students in a school went on a trip. How many students remained in school?
    (A) 27   (B) 30   (C) 873   (D) 773

31. One percent of $23,000 is
    (A) $23   (B) $2.30   (C) $230   (D) $2300

32. If a suit is on sale for $120, and the original cost was $150, what percentage would you save by buying it on sale?
    (A) 40 percent   (B) 30 percent   (C) 10 percent   (D) 20 percent

33. What is the percentage composition of oxygen in a mole of glucose ($C_6H_{12}O_6$)? (C = 12, H = 1, O = 16)
    (A) 53   (B) 35   (C) 6   (D) 20

34. A clerk is requested to file 800 cards. If he can file cards at the rate of 80 cards per hour, the number of cards remaining to be filed after seven hours of work is
    (A) 40   (B) 240   (C) 140   (D) 260

35. If a patient is required to get 45 minutes of exercise each day, how many hours of exercise does he get in a week?
    (A) 315 hours   (B) 5.25 hours   (C) 52 hours   (D) 6.4 hours

36. A truck going at a rate of 40 miles per hour will reach a town 80 miles away in how many hours?
    (A) 1 hour   (B) 3 hours   (C) 2 hours   (D) 4 hours

37. If $1,000 is the cost of repairing 100 square yards of pavement, the cost of repairing one square yard is
    (A) $10   (B) $150   (C) $100   (D) $300

38. If the average cost of sweeping a square foot of a small town's street is $0.75, the cost of sweeping 100 square feet is
    (A) $7.50   (B) $750   (C) $75   (D) $70

39. To work off 60 calories, Brenda needs to walk on the treadmill for 15 minutes. How long will it take her to work off 100 calories?
    (A) 25 minutes   (B) 55 minutes   (C) 40 minutes   (D) 45 minutes

40. If a kilogram equals about 35 ounces, the number of grams in 1 ounce is approximately
    (A) 29   (B) 30   (C) 31   (D) 32

41. An IV pump delivers medication at a constant rate of 24 milligrams per hour. How long does it take to deliver 90 milligrams?
    (A) 3 hours 15 minutes   (C) 3 hours 75 minutes
    (B) 3 hours 45 minutes   (D) 4 hours 15 minutes

42. If sound travels at the rate of 1,100 feet per second, in one-half minute it will travel about
    (A) 6 miles.   (B) 8 miles.   (C) 10 miles.   (D) 3 miles.

43. If a kilometer is about five eighths of a mile, 2 miles is about
    (A) 1.6 kilometers.   (B) 3.2 kilometers.   (C) 2.4 kilometers.   (D) 3.75 kilometers.

44. What is the cost of 5,500 bandages at $50 per thousand?
    (A) $385   (B) $550   (C) $275   (D) $285

45. At $1,250 per hundred, 228 watches will cost
    (A) $2,850   (B) $36,000   (C) $2,880   (D) $360

46. If cloth costs $42\frac{1}{2}$ cents per yard, how many yards can be purchased for $76.50?
    (A) 220   (B) 180   (C) 190   (D) 230

47. If 4 ounces of protein provide 448 calories, how much protein is needed to provide 392 calories?
    (A) 2.8 ounces   (B) 3.5 ounces   (C) 4.57 ounces   (D) 56 ounces

48. Men's handkerchiefs cost $1.29 for three. The cost per dozen handkerchiefs is
    (A) $7.74   (B) $3.87   (C) $14.48   (D) $5.16

49. 345 safety pins at $4.15 per hundred will cost
    (A) $0.1432   (B) $1.4320   (C) $14.32   (D) $143.20

50. A punch recipe for a half gallon (64 ounces) of punch requires one pint (16 ounces) of grape juice. How many quarts (1 quart = 32 ounces) of grape juice are required for $2\frac{1}{2}$ gallons of the punch?
    (A) 5 quarts   (B) 10 quarts   (C) $1\frac{1}{4}$ quarts   (D) $2\frac{1}{2}$ quarts

51. How many $\frac{3}{4}$ gram tablets are needed for a dosage of $4\frac{1}{2}$ grams?
    (A) 3.75   (B) 1.5   (C) 6   (D) 3

52. The oxidation of 1 gram of CHO produces 4 calories. How much CHO must be oxidized in the body to produce 36 calories? (CHO is a carbohydrate.)
    (A) 4 grams   (B) 7 grams   (C) 9 grams   (D) 12 grams

53. If there are 245 sections in a city containing five boroughs, the average number of sections for each of the five boroughs is
    (A) 50 sections.   (B) 49 sections.   (C) 47 sections.   (D) 59 sections.

54. If, in that same city, a section has 45 miles of street to plow after a snowstorm, and nine plows are used, each plow will cover an average of how many miles?
    (A) 7 miles   (B) 6 miles   (C) 8 miles   (D) 5 miles

55. If a crosswalk plow engine is run 5 minutes a day for ten days in a given month, how long will it run in the course of this month?
    (A) 50 minutes   (B) $1\frac{1}{2}$ hours   (C) 1 hour   (D) 30 minutes

56. If a sanitation department scow is towed at the rate of 3 miles per hour, how many hours, will it need to go 28 miles?
    (A) 10 hours 30 minutes   (C) 9 hours 20 minutes
    (B) 12 hours   (D) 9 hours 15 minutes

57. A cogwheel having eight cogs plays into another cogwheel having 24 cogs. When the small wheel has made 42 revolutions, how many has the larger wheel made?
    (A) 14   (B) 20   (C) 16   (D) 10

58. If $1\frac{1}{2}$ pounds of candy are required to fill an Easter basket, how many baskets can be filled with $10\frac{1}{2}$ pounds of candy?
    (A) 7.5   (B) 2.5   (C) $5\frac{1}{2}$ (D) 7

59. If 1 tee shirt costs $5.60, how many tee shirts can be bought for $61.60?
    (A) $8\frac{1}{2}$ (B) 10   (C) 9   (D) 11

60. The number of degrees on the Fahrenheit thermometer between the freezing point and the boiling point of water is
    (A) 100 degrees.   (B) 180 degrees.   (C) 212 degrees.   (D) 273 degrees.

Matter and The Periodic Table

SYNONYMS

Directions: In each of the sentences below, one word is in italics. Following each sentence are four words or phrases. For each sentence, select the word or phrase that best corresponds in meaning to the italicized word.

1. The chart *classifies* these organisms.
   (A) fuses   (B) categorizes   (C) controls   (D) camouflages

2. Biogenesis is *a theory* that supports that living things can be produced only from other living things.
   (A) a principle   (C) a rationale
   (B) an opinion   (D) an experimentation

ANTONYMS

Directions: For each of the following test items, select the word that is opposite in meaning to the term printed in capital letters.

3. HOMOGENOUS
   (A) invariable   (B) homosexual   (C) importunate   (D) heterogeneous

4. POSITIVE
   (A) negative   (B) sensitive   (C) diplomatic   (D) popular

5. DUCTILE
   (A) feted   (D) abnormal
   (B) alluvial   (E) belabor
   (C) stubborn

6. ABSORB
   (A) emit   (B) engulf   (C) engross   (D) consume

SPELLING USAGE

Directions: Select the letter that belongs in the blank space in the sentence.

7. The nurse had to _____ the baby.
   (A) weigh   (B) way

8. She wanted to purchase two _____ of milk
   (A) quartz   (B) quarts

Directions: Each question or incomplete statement below is followed by four suggested answers of completions.   Select the best answer choice.

9.   A calorie is a form of
    (A) light.   **(B)** heat.   **(C)** darkness.   **(D)** sound.

10.   Which of the following properties is considered a physical property?
    (A) flammability   **(B)** boiling point   **(C)** reactivity   **(D)** osmolarity

11.   Ozone is a molecular variety of
    (A) oxygen.   **(B)** chlorine.   **(C)** hydrogen.   **(D)** sulfur.

12.   Which of the following substances is a chemical compound?
    (A) blood   **(B)** water   **(C)** oxygen   **(D)** air

13.   The composition of edible sodium chloride (NaCl) from an explosive metal (Na) and a poisonous gas (Cl) illustrates which characteristics of matter?
    (A) impenetrability   **(C)** density
    (B) malleability   **(D)** emergent properties

Questions 14-20 refer to the periodic table below.

Periodic Table

IA																VIIA	Zero
H 1	IIA											IIIA	IVA	VA	VIA	H 1	He 2
Li 3	Be 4											B 5	C 6	N 7	O 8	F 9	Ne 10
Na 11	Mg 12	IIIB	IVB	VB	VIB	VIIB		VIII		IB	IIB	Al 13	Si 14	P 15	S 16	Cl 17	Ar 18
K 19	Ca 20	Sc 21	Ti 22	V 23	Cr 24	Mn 25	Fe 26	Co 27	Ni 28	Cu 29	Zn 30	Ga 31	Ge 32	As 33	Se 34	Br 35	Kr 36
Rb 37	Sr 38	Y 39	Zr 40	Nb 41	Mo 42	Tc 43	Ru 44	Rh 45	Pd 46	Ag 47	Cd 48	In 49	Sn 50	Sb 51	Te 52	I 53	Xe 54
Cs 55	Ba 56	*La 57	Hf 72	Ta 73	W 74	Re 75	Os 76	Ir 77	Pt 78	Au 79	Hg 80	Tl 81	Pb 82	Bi 83	Po 84	At 85	Rn 86
Fr 87	Ra 88	**Ac 89															

*Lanthanide Series	Ce 58	Pr 59	Nd 60	Pm 61	Sm 62	Eu 63	Gd 64	Tb 65	Dy 66	Ho 67	Fr 68	Tm 69	Yb 70	Lu 71
**Actinide Series	Th 90	Pa 91	U 92	Np 93	Pu 94	Am 95	Cm 96	Bk 97	Cf 98	Es 99	Fm 100	Md 101	No 102	Lr 103

14.   Which of the following elements is a transition element?
    (A) argon   **(B)** copper   **(C)** barium   **(D)** aluminum

15.   Which of the following is an example of a transition element?
    (A) aluminum   **(B)** astatine   **(C)** nickel   **(D)** rubidium

16. The elements in group Zero of the periodic table are considered inert gases because each has how many electrons in its outermost energy level (excluding He)?

   (A) 8   (B) 7   (C) 4   (D) 2

17. Identify the two atoms with the same number of electrons in their outermost energy level.

   (A) Na/K   (B) K/Ca   (C) Na/Mg   (D) Ca/Na

18. By use of the periodic table, it can be determined that the atoms with the greatest affinity would be

   (A) Na and Cl (B) K and F   (C) Na and F   (D) K and Cl

19. Using the periodic table, determine if the valence of a sodium ion is:

   (A) +1   (B) 0   (C) -1   (D) +12

20. Of the following groups the least reactive are

   (A) the halogens.   (C) group IIA metals.
   (B) the inert gases.   (D) precious metals of group IB.

21. The lightest element known on earth is

   (A) hydrogen. (B) helium.   (C) oxygen.   (D) air.

22. Which is an inert element?

   (A) hydrogen   (B) neon   (C) oxygen   (D) nitrogen

23. The element with the highest ionization energy of those below is

   (A) Mg   (B) Sr   (C) Ca   (D) Ba

24. The element with the smallest atomic radius of the following is

   (A) Sr   (B) Mg   (C) Ba   (D) Ra

25. The least electronegative of the following elements is

   (A) Cl   (B) F   (C) Br   (D) I

26. Which of the following arrangements gives the correct trend of electronegativism?

   (A) $I < Br < Cl < F$ (C) $Al > Si > P > S$
   (B) $Sr < Ca < Ra < Mg$ (D) $Na < K < Li < H$

27. The number of unpaired electrons in the outer subshell of a phosphorus atom (atomic number: 15) is

   (A) 2   (B) 0   (C) 3   (D) 1

28. An atom that has five 3 p electrons in its ground state is

   (A) Si   (B) P   (C) Cl   (D) O

29. How many valence electrons are needed to complete the outer valence shell of sulfur?

   (A) 1   (B) 2   (C) 3   (D) 4

30. An atom has the electron configuration (2-8-8-2). This atom would tend to

   (A) gain electrons.   (C) be inert.
   (B) lose electrons.   (D) None of the above.

31. The valence number of sulfur in the ion SO_4^{2-} is
    (A) -2   (B) +2   (C) +6   (D) +10

32. In sulfuric acid, the valence number of sulfur is
    (A) +2   (B) -2   (C) -4   (D) +6

33. The chemical name for sulfuric acid is
    (A) hydrogen sulfate   (C) sulfur trioxide
    (B) hydrogen sulfite   (D) hydrogen sulfide

34. The formula for sodium bisulfate is
    (A) $NaBiSO_4$ (B) $NaHSO_4$ (C) NaH_2SO_4 (D) Na_2SO_4

35. Which of the following species will combine with a chloride ion to produce ammonium chloride?
    (A) NH_3 (B) K^+ (C) NH_4^+ (D) Al^{+++}

36. A hydride and a hydrogen atom both have
    (A) the same number of electrons.   (C) the same number of protons.
    (B) the same charge.   (D) equal atomic radii.

37. The compound NaClO is called
    (A) sodium perchlorate.   (C) sodium chlorate.
    (B) sodium oxychloride.   (D) sodium hypochlorite.

38. The symbol for 2 molecules of hydrogen is
    (A) H_2 (B) $2H$ (C) $2H^+$ (D) $2H_2$

39. Commercial bleach has the formula
    (A) $CaCl_2$ (B) $Ca(OCl)_2$ (C) $CaOCl_2$ (D) $Ca(ClO_3)_2$

40. Two atoms have the same atomic number but different atomic weights (masses); therefore, these atoms are
    (A) compounds.   (C) neutrons.
    (B) isotopes.   (D) different elements.

41. The periodic table shows that the atomic number of fluorine is 9; this indicates that the fluorine atom contains
    (A) nine neutrons in its nucleus.
    (B) nine protons in its nucleus and nine electrons in orbit around the nucleus.
    (C) a total of nine protons and neutrons.
    (D) a total of nine protons and electrons.

42. What is the atomic weight of the element in the figure below?

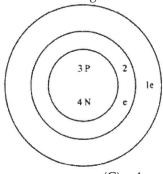

    (A) 2   (B) 3   (C) 4   (D) 7

43. If the atomic number of magnesium is 12, what will be the number of protons?
    (A) 6   (B) 10   (C) 14   (D) 12

44. Atoms are electrically neutral. This means that an atom will contain
    (A) more protons than neutrons.   (C) an equal number of protons and electrons
    (B) more electrons than protons.   (D) None of the above.

45. Atoms that have the same atomic number but different atomic masses
    (A) are from different elements.   (C) have different numbers of electrons.
    (B) are isobars.   (D) are isotopes.

46. Per atom, an element that has an atomic number of 19 contains
    (A) 19 electrons and 19 neutrons.   (C) 19 protons and 19 neutrons.
    (B) 19 electrons and 19 protons.   (D) a total of 19 protons and neutrons.

47. $^{34}_{17}Cl$ has
    (A) 17 protons, 17 electrons, and 17 neutrons.   (C) 17 protons, 18 electrons, and 17 neutrons.
    (B) 17 protons, 19 electrons, and 17 neutrons.   (D) 34 protons, 34 electrons, and 17 neutrons.

48. What is the gram molecular weight of $C_6H_{12}O_6$? (C = 12, H = 1, O = 16)
    (A) 29 grams   (B) 174 grams   (C) 180 grams   (D) 696 grams

49. What is the molecular weight of NaOH? (Atomic weights: Na = 23, O =16, H = 1)
    (A) 40 x 1   (B) 40 x 10   (C) 40 x 100   (D) 40 x 1,000

50. The percentage of oxygen by weight in $Al_2(SO_4)_3$ (atomic weights: Al = 27, S = 32, O = 16) is approximately
    (A) 19   (B) 21   (C) 56   (D) 92

51. For a molecular substance, a gram formula weight
    (A) is unrelated to the gram molecular weight.
    (B) is always equal to the mass corresponding to its empirical formula.
    (C) can always be calculated from its empirical formula alone.
    (D) is identical to its gram molecular weight.

Bonding, Structure, and Reactions

Directions: Each question or incomplete statement below is followed by four suggested answers of completions. Select the best answer choice.

1. In forming an ionic bond with an atom of chlorine, a sodium atom will
   - **(A)** receive 1 electron from the chlorine atom.
   - **(B)** receive 2 electrons from the chlorine atom.
   - **(C)** give up 1 electron to the chlorine atom.
   - **(D)** give up 2 electrons to the chlorine atom

2. The portion of an atom directly involved in the ionic bonding is/are
   - **(A)** protons in the nucleus.
   - **(B)** neutrons in the nucleus.
   - **(C)** electrons in the outer energy level.
   - **(D)** electrons in the innermost energy level.

3. A covalent bond is believe to be caused by
   - **(A)** transfer of electrons.
   - **(B)** sharing of electrons.
   - **(C)** release of energy.
   - **(D)** None of the above.

4. Polar bonds form when
   - **(A)** electrons are shared unequally between two atoms.
   - **(B)** more than one pair of electrons is shared.
   - **(C)** ions are formed.
   - **(D)** an acid and base are combined.

5. Which of the following is an example of hydrogen bonding?
   - **(A)** the bond between O and H in a single molecule of water
   - **(B)** the bond between O of one water molecule and H of a second water molecule
   - **(C)** the bond between O of one water molecule and the O of a second water molecule
   - **(D)** the bond between H of one water molecule and H of a second water molecule

6. When calcium reacts with chlorine to form calcium chloride, it
   - **(A)** shares two electrons.
   - **(B)** gains two electrons.
   - **(C)** loses two electrons.
   - **(D)** gains one electron.

7. Which statement is incorrect?
   - **(A)** London dispersion forces are among those binding the units of molecular solids.
   - **(B)** Molten ionic compounds are conductors of electricity.
   - **(C)** Molecular solids have high melting points.
   - **(D)** Molecular solids are non-conductors.

8. Which of the following groups contains no ionic compounds?
   - **(A)** HCN, NO, $Ca(NO_3)_2$
   - **(B)** KOH, CCl_4, SF_6
   - **(C)** NaH, CaF_2, $NaNH_2$
   - **(D)** CH_2O, H_2S, NH_3

9. An ion is
   - **(A)** one molecule of water.
   - **(B)** one particle of hydrogen.
   - **(C)** the same as a neutron.
   - **(D)** an atom with an electric charge.

ANTONYM

Directions: For the following test item, select the word that is opposite in meaning to the term printed in capital letters.

10. SYMMETRY
    (A) invocation
    (B) madrigal
    (C) distortion
    (D) satyr
    (E) cilia

11. In a cubic lattice, an atom lying at the corner of a unit cell is shared by how many unit cells?
    (A) 2
    (B) 4
    (C) 8
    (D) 12

12. Bronze is an alloy of copper and
    (A) iron.
    (B) lead.
    (C) zinc.
    (D) tin.

13. Oxygen gas can be obtained in appreciable quantities by heating all of the following EXCEPT
    (A) H_2O
    (B) H_2O_2
    (C) HgO
    (D) PbO_2

14. The chemical reaction $2 Zn + 2 HCl \rightarrow 2 ZnCl + H_2$ is an example of
    (A) double displacement.
    (B) synthesis.
    (C) analysis.
    (D) single displacement.

15. A solution of zinc chloride should NOT be stored in a tank made of aluminum because
    (A) aluminum will displace the zinc in the zinc chloride solution.
    (B) the zinc will become contaminated.
    (C) the chloride ion will react with impurities in the solution.
    (D) the two metals will react to produce an undesirable compound.

16. The complete combustion of carbon disulfide would yield carbon dioxide and
    (A) sulfur.
    (B) sulfur dioxide.
    (C) sulfuric acid.
    (D) water.

17. Of the following compounds, which is more difficult to decompose than lithium fluoride?
    (A) lithium bromide
    (B) lithium chloride
    (C) lithium iodide
    (D) None of the above.

18. When copper oxide is heated with charcoal, the reaction that occurs is an example of
    (A) reduction only.
    (B) oxidation only.
    (C) oxidation and reduction.
    (D) neither oxidation nor reduction.

19. There is no oxidation-reduction in a reaction that involves
    (A) single replacement.
    (B) double replacement.
    (C) simple decomposition.
    (D) direct combination of elements.

20. Which of the following equations represents an oxidation-reduction reaction?
    (A) $2 Na + Cl_2 \rightarrow 2 NaCl$
    (B) $CO_2 + H_2O \rightarrow H_2CO_3$
    (C) $HNO_3 + KOH \rightarrow KNO_3 + H_2O$
    (D) $CaO + H_2O \rightarrow Ca(OH)_2$

21. The oxidation number of Mn in the compound $KMnO_4$ is
    (A) +7   (B) +2   (C) 0   (D) +6

22. The best reducing agent is
    (A) mercury   (B) hydrogen   (C) copper   (D) carbon dioxide

23. All of the reactions between the following pairs will produce hydrogen EXCEPT
    (A) copper and hydrochloric acid.
    (B) iron and sulfuric acid.
    (C) magnesium and steam.
    (D) sodium and alcohol.

24. The gas resulting when hydrochloric acid is added to a mixture of iron filings and sulfur is
    (A) H_2S (B) SO_2 (C) SO_3 (D) H_2

25. When carbon dioxide gas is bubbled through water in a test tube, the product is
    (A) ozone.
    (B) methane.
    (C) hydrogen peroxide.
    (D) carbonic acid.

26. When hydrochloric acid is added to sodium sulfite, and the gas that is formed is bubbled through barium hydroxide, the salt formed is
    (A) $BaCl_2$ (B) $BaSO_3$ (C) NaCl   (D) NaOH

27. Which one of the following equations represents neutralization?
    (A) $2\,Na + Cl_2 \rightarrow 2\,NaCl$
    (B) $CO_2 + H_2O \rightarrow H_2CO_3$
    (C) $HNO_3 + KOH \rightarrow KNO_3 + H_2O$
    (D) $CaO + H_2O \rightarrow Ca(OH)_2$

28. Identify the products of the chemical reaction between 1 cubic centimeter of KOH and 1 cubic centimeter of HCl.
    (A) $K^+ + Cl_2 + 2\,H + OH$
    (B) $KCl + H_2 + O_2$
    (C) $K + Cl + HOH$
    (D) $KCl + H_2O$

29. Which of the following could react chemically with ammonia?
    (A) H   (B) Cl^- (C) Na   (D) H^+

30. Which of the following salts would be more soluble in 1.0 M acid than in pure water?
    (A) KCl   (B) $CaCO_3$ (C) $CaCl_2$ (D) KNO_3

31. An emission device in modern cars that uses platinum beads to oxidize carbon monoxide and hydrocarbons to carbon dioxide and water is the
    (A) carburetor.
    (B) catalytic converter.
    (C) PCV valve.
    (D) air filter.

32. Which one of the following equations is balanced?
    (A) $H_2O \rightarrow H_2 \uparrow + O_2 \uparrow$
    (B) $Al + H_2SO_4 \rightarrow Al_2(SO_4)_3 + H_2 \uparrow$
    (C) $S + O_2 \rightarrow SO_3$
    (D) $2\,HgO \rightarrow 2\,Hg + O_2 \uparrow$

33. If two moles of compound A react with 5 moles of compound B to form compound C, then how many moles of A are required to react completely with 7 moles of B to form compound C?
    (A) 5.7   (B) 2.8   (C) 7.5   (D) 8.2

34. The number of grams of hydrogen formed by the action of 6 grams of magnesium (atomic weight = 24) on an appropriate quantity of acid is
    (A) 0.5   (B) 8   (C) 22.4   (D) 72

35. If a eudiometer tube is filled with 26 milliliters of hydrogen and 24 milliliters of oxygen and the mixture exploded, which of the following would remain uncombined?
    (A) 2 milliliters hydrogen   (C) 23 milliliters hydrogen
    (B) 14 milliliters hydrogen   (D) 11 milliliters oxygen

36. In a titration of 40.0 mL of 0.20 M NaOH with 0.4 M HCl, what will be the final volume of the solution when the sodium hydroxide is completely neutralized?
    (A) 42 mL   (B) 20 mL   (C) 60 mL   (D) 80 mL

37. A compound that has a high heat of formation is normally
    (A) easy to form from its elements and easy to decompose.
    (B) easy to form from its elements and difficult to decompose.
    (C) difficult to form from its elements and easy to decompose.
    (D) difficult to form from its elements and difficult to decompose.

38. In exergonic reactions, the energy is
    (A) used.   (B) stored.   (C) released.   (D) lost.

39. Identify the statement which is NOT characteristic of exergonic reactions.
    (A) They are downhill reactions.
    (B) They have a negative energy change (-H).
    (C) They are uphill reactions.
    (D) The products have less energy than the reactants.

40. The second law of thermodynamics (entropy) states that systems (ranging from a single organism to the entire universe) become increasingly disordered or random with time. Early scientists erroneously believed that the human body, by maintaining its structural and functional integrity violated this law. It is currently understood that the human body conforms to the second law of thermodynamics because it is a(an)
    (A) closed system.
    (B) assembly of organic molecules.
    (C) open system.
    (D) composite of inorganic and organic molecules.

41. Reaction kinetics deals with
    (A) equilibrium position.   (C) molecular reactant size.
    (B) reaction rates.   (D) None of the above.

42. If the reaction: $A + B \rightarrow C + D$ is designated as first order, the rate depends on
    (A) the concentration of only one reactant.   (C) no specific concentration.
    (B) the concentration of each reactant.   (D) the temperature only.

Questions 43-45 refer to the following graph and equation.

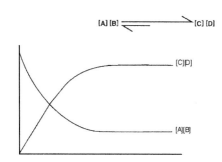

43. At time zero, the percentage of substance that is [A] [B] is approximately
    (A)   100   (B)   0   (C)   50   (D)   25

44. The comparative affinity of the reactants and products is
    (A)   [C] [D] > [A] [B]   (C)   [A] [B] > [C] [D]
    (B)   [A] [B] = [C] [D]   (D)   None of the above

45. Choose the incorrect statement, relative to the reaction at equilibrium.
    (A)   Concentration of reactants and products remains constant.
    (B)   Reaction is shifted to the right.
    (C)   Forward and reverse reactions occur at equal rates.
    (D)   Reaction is shifted to the left.

46. In a different chemical reaction, [A] and [B] combine to form [C] and [D], as expressed by the reaction
    [A][B]=[C][D]. Select the statement that best describes the equilibrium condition.
    (A)   The reaction is shifted to the right.
    (B)   Concentration of reactants and products are constant.
    (C)   Reaction is shifted to the left.
    (D)   Concentration of products is greater than the concentration of reactants.

47. For the reaction: H_2 (g) + Br_2 (g) $\rightleftharpoons$ 2 HBr (g), the reaction can be driven to the left by
    (A)   increasing the pressure.   (C)   increasing the hydrogen bromide.
    (B)   increasing the hydrogen.   (D)   decreasing the hydrogen bromide.

48. Consider the reaction N_2 (g) + 3 H_2 (g) $\rightleftharpoons$ 2 NH_3 (g) + heat. Indicate the incorrect statement.
    (A)   An increase in temperature will shift the equilibrium to the right.
    (B)   An increase in pressure applied will shift the equilibrium to the right.
    (C)   The addition of ammonia will shift the equilibrium to the left.
    (D)   The addition of H_2 will shift the equilibrium to the right.

Gases, Boiling, and Specific Heat

Directions: Each question or incomplete statement below is followed by four suggested answers of completions. Select the best answer choice.

1. A gas lighter than air is
   (A) CH_4
   (B) C_6H_6
   (C) HCl
   (D) N_2O

2. Of the following gases, which is odorless and heavier than air?
   (A) CO
   (B) CO_2
   (C) H_2S
   (D) N_2

3. The instrument used to measure air pressure is called a
   (A) thermometer.
   (B) hydrometer.
   (C) barometer.
   (D) sphygmomanometer.

4. The normal height of a mercury barometer at sea level is
   (A) 15 inches.
   (B) 30 inches.
   (C) 32 feet.
   (D) 34 feet.

5. The most abundant gas in the atmosphere is
   (A) oxygen.
   (B) nitrogen.
   (C) carbon dioxide.
   (D) chlorine.

6. Of the following gases in the air, the most plentiful is
   (A) argon.
   (B) nitrogen.
   (C) oxygen.
   (D) air.

7. Of the following, the one present in greatest amounts in dry air is
   (A) carbon dioxide.
   (B) oxygen.
   (C) water vapor.
   (D) nitrogen.

8. The nurse is caring for a patient who is receiving 90% oxygen. Before putting the mask on, he checks the oxygen analyzer to determine that it is delivering the correct amount. What percentage of oxygen should the analyzer indicate when it is in room air?
   (A) 20 percent
   (B) 90 percent
   (C) 50 percent
   (D) 8 percent

9. In a volume of air at pressure of one atmosphere at sea level, the partial pressure of oxygen is equal to
   (A) 593 mm of mercury.
   (B) 494 mm of mercury.
   (C) 380 mm of mercury.
   (D) 160 mm of mercury.

10. In a volume of air at one atmosphere pressure at sea level, the partial pressure of nitrogen will be about
    (A) 490 mm of mercury.
    (B) 760 mm of mercury.
    (C) 106 mm of mercury.
    (D) 608 mm of mercury.

11. Evaporation of water is likely to be greatest on days of
    (A) high humidity.
    (B) low humidity.
    (C) little or no wind.
    (D) low pressure.

12. In order to increase the temperature of a gas in a closed unit, it would be necessary to
    (A) increase the pressure.
    (B) decrease the density.
    (C) decrease the volume.
    (D) increase the space.

13. Which one of the following graphs represents Boyle's Law?

14. The inhaling and exhaling of air by the human lungs is mainly an application of
    (A) Boyle's Law – the inverse relationship between the pressure and the volume of a gas.
    (B) the volume of a gas at standard temperature and pressure (STP).
    (C) Charles' Law – the direct relationship between the temperature and the volume of a gas.
    (D) the number of O_2 and CO_2 particles per mole.

15. Two parameters (e.g., the volume and the temperature of a gas) are directly proportional if a constant value can be calculated from their
    (A) product.   (B) ratio.   (C) sum.   (D) difference.

16. Consider three 1-liter flasks at STP. Flask A contains NO gas; Flask B contains NH_3 gas; and Flask C contains N_2 gas. Which flask contains the greatest number of atoms?
    (A) Flask A   (C) Flask C
    (B) Flask B   (D) All contain the same number of atoms.

17. The weight in grams of 22.4 liters of nitrogen (atomic weight = 14) is
    (A) 3   (B) 7   (C) 14   (D) 28

18. One liter of a certain gas, under standard conditions, weighs 1.16 g. A possible formula for the gas is
    (A) C_2H_2 (B) CO   (C) NH_3 (D) O_2

19. If gas A has a molecular weight four times that of gas B,
    (A) the average speed of gas A is about 4 times that of gas B.
    (B) the average speed of gas B is about 4 times that of gas A.
    (C) the average speed of gas A is about twice that of gas B.
    (D) the average speed of gas B is about twice that of gas A.

20. What are the differentiating factors for potential and kinetic energy?
    (A) Properties – physical or chemical
    (B) State – solid or liquid
    (C) Temperature – high or low
    (D) Activity – in motion or storage

21. Stored energy is referred to as
    (A) activation energy.   (C) potential energy.
    (B) kinetic energy.   (D) electrical energy.

22. On the basis of the following boiling point data, which of the following liquids would be expected to have the highest vapor pressure at room temperature?

	Substance	Boiling Point		Substance	Boiling Point
(A)	acetone	56.2°C	(C)	water	100°C
(B)	ethanol	78.5°C	(D)	ethylene glycol	198°C

23. The boiling points of the following gases are as follows: argon, -185.7°C; helium, -268.9°C; nitrogen, -195.8°C; oxygen, -183°C. In the fractional distillation of liquid air, the gas that boils off last is
    (A) argon.   **(B)** helium.   **(C)** nitrogen.   **(D)** oxygen.

24. The reason why concentrated H_2SO_4 is used extensively to prepare other acids is that concentrated sulfuric acid
    (A) is highly ionized.   **(C)** has a high specific gravity.
    (B) is an excellent dehydrating agent.   **(D)** has a high boiling point.

25. If 3,480 calories of heat are required to raise the temperature of 300 grams of a substance from 50°C to 70°C, the substance would be: Use the formula:
    $$c = Q/MDT$$
    c = specific heat
    Q = number of calories
    M = mass
    T = temperature

Substance	Specific Heat		Substance	Specific Heat
(A) ethyl alcohol	0.581	**(C)**	liquid ammonia	1.125
(B) aluminum	0.214	**(D)**	water	1.0

26. How many Calories are required to change the temperature of 2,000 grams of H_2O from 20°C to 38°C?
    (A) 36 Calories   **(B)** 24 Calories   **(C)** 18 Calories   **(D)** 12 Calories

Solutions

SYNONYMS

Directions: In the sentence below, one word is in italics. Following the sentence are four words or phrases. Select the word or phrase that best corresponds in meaning to the italicized word.

1. *Saturated* with moisture
   (A) void of   (B) mixed with   (C) full of   (D) replaced with

Directions: Each question or incomplete statement below is followed by four suggested answers of completions. Select the best answer choice.

2. When you mix salt with water, what is the water called?
   (A) solute   (B) solvent   (C) solution   (D) ionizer

3. A solution that contains all the solute it can normally dissolve at a given temperature must be
   (A) concentrated.   (B) supersaturated.   (C) saturated.   (D) unsaturated.

4. If the stirring of a solution results in precipitation of solute with no change in temperature, the solution must have been
   (A) saturated.   (B) concentrated.   (C) dilute.   (D) supersaturated.

5. As water is warmed, its solubility of oxygen (O_2)
   (A) increases.   (C) remains constant.
   (B) decreases.   (D) fluctuates randomly.

6. As you try to mix water and oil in your salad dressing, they do not mix because
   (A) water is hydrophilic and oil is hydrophobic.
   (B) water is polar and oil is nonpolar.
   (C) both are hydrophilic.
   (D) None of the above.

7. A 10-percent solution of glucose will contain
   (A) 1 gram of glucose per 1,000 milliliters of solution.
   (B) 1 gram of glucose per 100 milliliters of solution.
   (C) 1 gram of glucose per 10 microliters of solution.
   (D) 10 grams of glucose per 100 milliliters of solution.

Question 8 refers to the diagram below.

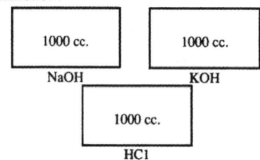

Each diagram represents one solution: One gram molecular weight of NaOH, one of KOH, and one of HCl, each dissolved in enough H_2O to make 1 liter.

8. These are molar quantities because
    (A) their molecular weights are equal to each other.
    (B) the volume for each solution is the same.
    (C) each solution contains 1 gram molecular weight.
    (D) the percentage of solute to solvent is equal in each solution.

9. A 1-molar solution of K_3PO_4, potassium phosphate, contains in 1 liter
    (A) one mole of potassium ions.
    (C) one mole of phosphorus atoms.
    (B) one mole of oxygen atoms.
    (D) no ions.

10. Per liter, compared to a 3 molar aqueous solution, a 3 molal aqueous solution contains
    (A) the same amount of solute.
    (C) less solute.
    (B) more solute.
    (D) a variable amount of solute.

11. For a solution of H_2SO_4, the equivalency between molarity and normality would be ($H = 1$, $S = 32$, $O = 16$)
    (A) $1 M = 1 N$ (B) $1 M = 2 N$ (C) $1 N = 2 M$ (D) $1 M = 0.5 N$

12. A nurse administered a medication at 10^{-4} molar concentrated to a patient. The doctor instructed that the dosage be reduced by one-half strength, which was
    (A) 10^{-2} M   (B) 5×10^{-5} M   (C) 10^{-8} M   (D) 10^{-3} M

13. The non-electrolyte in the following group is
    (A) acetic acid   (B) calcium chloride.   (C) sodium bromide.   (D) sugar.

14. An example of a strong electrolyte is
    (A) sugar.   (B) calcium chloride.   (C) glycerine.   (D) boric acid.

15. Ice can be melted most effectively by _____ if 1 mole is used.
    (A) sucrose   (B) calcium chloride   (C) sodium chloride   (D) methanol

16. All of the following are colligative properties of solution EXCEPT
    (A) vapor pressure.
    (C) density.
    (B) osmotic pressure.
    (D) boiling point elevation.

17. If the osmotic pressure of human blood is determined by the number of dissolved particles, and a 0.315 molar glucose solution is isotonic to blood, the concentration of isotonic sodium chloride (NaCl) would be
    (A) 0.630 M   (B) 0.3 M   (C) 0.157 M   (D) 0.2 M

18. Cell membranes control the movement of substances into and out of the cell and are best described as
    (A) selectively permeable.   (C) impermeable.
    (B) freely permeable.   (D) totally permeable.

19. The AIDS virus is transported in bodily fluids. The Surgeon General of the United States sends out information about the disease and its transmission. In one section, there is a recommendation that one should use latex (a form of plastic) condoms rather than those made of natural membranes. This recommendation is probably based upon the principal of
    (A) diffusion.
    (B) facilitated transport.
    (C) active transport.
    (D) varied selectivity, permeability of membranes.

20. The movement of substances from lesser concentration to higher concentration is called
    (A) osmosis.   (B) diffusion.   (C) active transport.   (D) pinocytosis.

21. Which term denotes the movement of glucose molecules from an area of lower concentration to an area of higher concentration?
    (A) osmosis   (B) diffusion   (C) dialysis   (D) active transport

22. Passage of water through the membrane of a cell is called
    (A) assimilation.   (B) osmosis.   (C) circulation.   (D) transpiration.

23. The diffusion of water through a semipermeable membrane is known as
    (A) anabolism.   (B) synthesis.   (C) mitosis.   (D) osmosis.

24. Cells that contain more dissolved salts and sugars than the surrounding solution are called
    (A) isotonic.   (B) hypertonic.   (C) hypotonic.   (D) osmosis.

25. If a red blood cell is placed in sea water, it will be in what kind of solution?
    (A) isotonic   (C) hypertonic
    (B) hypotonic   (D) facilitated diffusion

26. If 5×10^5 lbs of NaCl were dumped into a 0.5 acre pond, what would happen to the water concentration inside the frog's body cells?
    (A) It would not change.   (C) It would decrease.
    (B) It would increase.   (D) It would approach boiling.

27. Mammalian cells suspended within a hypertonic solution will
    (A) swell and possibly rupture.
    (B) lose water by osmosis.
    (C) take on water by osmosis.
    (D) remain unchanged since intracellular fluid is also hypertonic.

28. You place a cell in a solution of substance x and water. Substance x is always present in the cell, but you do not know the concentration ratio in either case. The cell increases in size. What is the tonicity of the solution in which you placed the cell?

(A) hypotonic (B) isotonic (C) hypertonic (D) None of the above

Acids, Bases, and Salts

Directions: Each question or incomplete statement below is followed by four suggested answers of completions. Select the best answer choice.

1. The formula that represents the strongest acid in the following group is
   - (A) HCl
   - (B) HCN
   - (C) HNO_2
   - (D) HCOOH

2. The order of strength of the following bases is: $HCO_3^- > C_2H_3O_2^- > HSO_4^- > Cl^-$, the weakest acid is
   - (A) $HCl\left(H^+ + Cl^-\right)$
   - (B) $HC_2H_3O_2\left(H^+ + C_2H_3O_2^-\right)$
   - (C) $H_2SO_4\left(H^+ + HSO_4^-\right)$
   - (D) $H_2CO_3\left(H^+ + HCO_3^-\right)$

3. Which of the following is NOT an acid/conjugate base pair?
   - (A) HCN/CN^-
   - (B) H_2CO_3/OH^-
   - (C) H_2SO_4/HSO_4^-
   - (D) $H_3PO_4/H_2PO_4^-$

4. A high concentration of H^+ ions is characteristic of
   - (A) high pH.
   - (B) strong acid.
   - (C) alkaline base.
   - (D) both (A) and (C).

5. In a 0.001 M solution of HCl, the pH is:
   - (A) 2
   - (B) -3
   - (C) 1
   - (D) 3

6. When a solution has a pH of 7, it is
   - (A) a strong base.
   - (B) a strong acid.
   - (C) a weak base.
   - (D) neutral.

7. A common detergent has pH 11.0, so the detergent is
   - (A) neutral.
   - (B) acidic.
   - (C) alkaline.
   - (D) None of the above.

8. Which is true of alkaline solutions?
   - (A) More H^+ than OH^- ion
   - (B) Same amount of H^+ ion and OH^- ion
   - (C) More OH^- ion than H^+ ion
   - (D) None of the above

9. The compound that will not be acidic when dissolved in water is
   - (A) HBr
   - (B) N_2O_5
   - (C) CaO
   - (D) NH_4Cl

10. An oxide whose water solution will turn litmus red is
    - (A) BaO
    - (B) Na_2O
    - (C) P_2O_3
    - (D) CaO

11. Nonmetal oxides, when dissolved in water, tend to form
    - (A) acids.
    - (B) bases.
    - (C) salts.
    - (D) hydrides.

12. Which of the following substances will raise the pH of a solution of hydrochloric acid?
    - (A) NaCl
    - (B) H_2CO_3
    - (C) $NaHCO_3$
    - (D) HNO_3

13. Which of the following compounds would be classified as a salt?
    - (A) Na_2CO_3
    - (B) $Ca(OH)_2$
    - (C) H_2CO_3
    - (D) CH_3OH

14. When dissolved in water to form 0.2 M solutions, which of the following would have the highest pH?
    (A) the salt of a strong acid
    (B) a weak acid
    (C) the ammonium salt of a strong acid
    (D) the sodium salt of a weak acid

15. The major benefit of buffer systems is that they
    (A) increase pH significantly.
    (B) resist significant changes in pH.
    (C) decrease pH significantly.
    (D) None of the above.

Nuclear Chemistry

SYNONYMS

Directions: In the sentence below, one word is in italics. Following the sentence are four words or phrases. Select the word or phrase that best corresponds in meaning to the italicized word.

1. Various courses were *fused* in the revision of the curriculum.
   (A) required   (B) implicated   (C) combined   (D) involved

Directions: Each question or incomplete statement below is followed by four suggested answers of completions. Select the best answer choice.

2. Which of the following is NOT a form of radioactive decay?
   (A) electron capture   (C) alpha emission
   (B) beta emission   (D) proton emission

3. Which of the following kinds of radiation is most penetrating?
   (A) Alpha   (B) Beta   (C) Gamma   (D) X-rays

4. The loss of an alpha particle from the radioactive atom $^{228}_{88}\text{Ra}$ would leave
   (A) $^{224}_{86}\text{Rn}$ (B) $^{222}_{86}\text{Rn}$ (C) $^{224}_{88}\text{Ra}$ (D) $^{230}_{90}\text{Th}$

5. Calculate the time required for 100 milligrams of ^{131}I (dissociation constant, K = 0.086625) to decay to 50 milligrams. Use the formula $K = \dfrac{0.693}{t_{\frac{1}{2}}}$.

   (A) 0.5 days   (B) 0.4 days   (C) 64 days   (D) 8 days

6. ^{131}I has a half-life of eight days. A 100-milligram sample of this radioactive element would decay to what amount after eight days.
   (A) 50 milligrams   (B) 40 milligrams   (C) 30 milligrams   (D) 20 milligrams

7. If a radioactive element with a half-life ($t_{\frac{1}{2}}$) of 100 years has 31.5 kg remaining after 400 years of decay, the amount in the original sample was close to
   (A) 2,500 kg.   (B) 500 kg.   (C) 50 kg.   (D) 5,000 kg.

8. Carbon-14 has a half-life of 5.73 x 10 years. If a sample contained 1 gram of C-14, the time required to decay to only 0.0625 g would be
   (A) 11.46 × 10 years   (C) 22.92 × 10 years
   (B) 5.73 × 10 years   (D) None of the above.

9. The fundamental principle expressed by the Einstein Equation ($E=mc^2$) on mass-energy equivalency is that
   (A) small mass = much energy.   (C) little energy = great mass.
   (B) small mass = little energy.   (D) All of the above.

Organic Chemistry

Directions: Each question or incomplete statement below is followed by four suggested answers of completions. Select the best answer choice.

1. An example of an organic compound is
   (A) water (H_2O).
   (B) ammonia (NH_3).
   (C) salt (NaCl).
   (D) glucose ($C_6H_{12}O_6$).

2. The generic formula for alkane, aliphatic hydrocarbons is C_nH_{2n+2}, from which the formula for propane is derived, is
   (A) C_2H_6
   (B) C_3H_8
   (C) NaNO
   (D) C_3H_4

3. Of the following, which is an aromatic compound?
   (A) benzene
   (B) ethyl alcohol
   (C) iodoform
   (D) methane

4. In the compound propene, $H_2C=CH-CH_3$, the single bond between two carbon atoms is
   (A) stronger than the double bond.
   (B) shorter than the double bond.
   (C) equal to the double bond in bond strength.
   (D) longer than the double bond.

5. Choose the correct structural formula for acetylene, C_2H_2.
   (A) HC=CH
   (B) HC-CH
   (C) HC ≡CH
   (D) HC≡≡≡CH

6. The general formula for the acetylene series of hydrocarbons is
   (A) C_nH_{2n+2}
   (B) C_nH_{2n}
   (C) C_nH_{2n-2}
   (D) None of the above

7. Alcoholic beverages contain
   (A) wood alcohol.
   (B) isopropyl alcohol.
   (C) glyceryl alcohol.
   (D) ethyl alcohol.

8. Rubbing alcohol is
   (A) methyl alcohol.
   (B) ethyl alcohol.
   (C) phenol.
   (D) isopropyl alcohol.

9. Which of the following compounds is an ether?
   (A) CH_3CHO
   (B) CH_3-O-CH_3
   (C) $CH_3--COOH$
   (D) CH_3-CH_2OH

10. The compound that has the greatest polarity is
    (A) $CH_3-CH_2-O-CH_2-CH_3$
    (B) $CH_3-CH_2CH_2CH_2-CH_3$
    (C) $CH_3-CH_2-CH_2-CH_2-CH_2-Cl$
    (D) $CH_3-CH_2-CH_2-CH_2-CH_2--OH$

11. The general formula for an aldehyde is
    (A) RCOOH
    (B) RCOOR
    (C) ROH
    (D) RCHO

12. Which of the following molecules would be classified as a ketone?

(A)

```
    H  H  H  H
    |  |  |  |
H — C — C — C — C = O
    |  |  |
    H  H  H
```

(C)

```
    H  H     H  H
    |  |     |  |
H — C — C — C — C — C — H
    |  |  ||  |  |
    H  H  O  H  H
```

(B)

```
    H  H  H  H  H
    |  |  |  |  |
H — C — C — C — C — C — H
    |  |  |  |  |
    H  H  H  OH H
```

(D)

```
    H  H  H  H  H
    |  |  |  |  |
H — C — C — C — C — C = O
    |  |  |  |
    H  H  OH H
```

13. The general formula for an organic acid is

(A) RCOOR   (B) ROH   (C) ROR   (D) RCOOH

Question 14 refers to the molecule below.

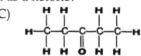

14. The above compound is a(an)

(A) amino acid.   (B) aldehyde.   (C) alpha-keto acid.   (D) alcohol.

Carbohydrates

Directions: Each question or incomplete statement below is followed by four suggested answers of completions. Select the best answer choice.

1. The basic building block of carbohydrate is
   (A) starch.   (B) chitin.   (C) sucrose.   (D) glucose.

2. Carbohydrates are a combination of carbon, hydrogen, and oxygen in an approximate ratio of
   (A) 2:1:2   (B) 3:2:1   (C) 1:2:1   (D) 1:1:1

3. Inasmuch as the molecular formula for glucose is $C_6H_{12}O_6$ and the molecular formula for fructose is $C_6H_{12}O_6$, the two substances are
   (A) hextomers.   (B) isomers.   (C) heteromers.   (D) anomers.

4. Of the following, which is a monosaccharide?
   (A) dextrose   (B) glycogen   (C) lactose   (D) sucrose

5. Which of the following is a dissaccharide?
   (A) glucose   (B) maltose   (C) fructose   (D) chilin

6. The sugar with the highest molecular weight of those listed is
   (A) fructose.
   (B) sucrose.
   (C) glucose.
   (D) None of the above; all have the same molecular weight.

7. Long chains of glucose molecules are involved in the structure of
   (A) proteins.   (B) fats.   (C) cholesterol.   (D) polysaccharides.

8. Another name for animal starch is
   (A) cellulose.   (B) lecithin.   (C) glycogen.   (D) chitin.

9. Which one of the following is NOT a carbohydrate?
   (A) maltose   (B) cellulose   (C) glycogen   (D) cholesterol

10. Which one of the following is NOT a carbohydrate?
    (A) maltose   (B) cellulose   (C) glycogen   (D) wax

11. When catalyzed by sucrase, sucrose decomposes to yield glucose + fructose. The reaction is
    (A) fermentation.   (B) hydrolysis.   (C) denaturation.   (D) condensation.

12. Glucose is a reducing sugar, which if boiled in Benedict's reagent, produces an orange to brick-red color. Choose the chemical species it reduces.
    (A) OH^- (B) Cu^{2+} (C) Cu (D) Cu^{2-}

Lipids

SYNONYMS

Directions: In the sentence below, one word is in italics. Following the are four words or phrases. For the sentence, select the word or phrase that best corresponds in meaning to the italicized word.

1. *Lipids* are organic compounds that store and release large amounts of energy.

   (A) enzymes   (B) diatoms   (C) proteins   (D) fats

Directions: Each question or incomplete statement below is followed by four suggested answers of completions. Select the best answer choice.

2. Oil and water are immiscible (do not mix) because
   (A) oil is polar and water is polar.   (C) water is nonpolar and oil is polar.
   (B) oil is nonpolar and water is polar.   (D) water is nonpolar and oil is nonpolar.

3. What happens when a small amount of soap is added to hard water?
   (A) Copious suds are formed.
   (B) A scum is formed.
   (C) All sediments filter out.
   (D) Sediments diffuse equally throughout the solution.

4. Foods containing unsaturated fatty acids are more healthy than those with saturated ones because they contain
   (A) more hydrogen.   (C) more nitrogen.
   (B) less oxygen.   (D) less hydrogen.

5. Fats belong to the class of organic compounds represented by the general formula, RCOOR', where R and R' represent hydrocarbon groups; therefore, fats are
   (A) ethers.   (B) soaps.   (C) esters.   (D) lipases.

6. Which of the following hormones is predominant in females?
   (A) Androgen   (B) Testosterone   (C) Gonadotrophin   (D) Estrogen

7. The hormone produced by the adrenal glands is
   (A) progesterone.   (B) estrogen.   (C) testosterone.   (D) aldosterone.

8. The ovaries produce the hormones
   (A) estrogen and testosterone.   (C) estrogen and progesterone.
   (B) progesterone and testosterone.   (D) progesterone and prolactin.

9. The hormone produced by the testes is
   (A) progesterone.   (B) estrogen.   (C) testosterone.   (D) aldosterone.

10. In females, _____ promotes estrogen secretion from cells within the ovaries, while in males, it stimulates sperm production within the testes.

(A) testosterone

(C) follicle stimulating hormone

(B) melanotropin

(D) luteinizing hormone

11. _____ stimulates the smooth muscle of the uterine wall during the labor and delivery process. After delivery, it promotes the ejection of milk.

(A) Oxytocin   (B) Vasopressin   (C) Somatotropin   (D) Prolactin

12. Cholesterol, in spite of its bad reputation, is an essential component of

(A) microtubules.

(C) ribosomes.

(B) the cell membrane.

(D) cytosol.

13. The basic structure of cell membranes is a

(A) protein bilayer.

(B) protein-impregnated phospholipid bilayer.

(C) carbohydrate bilayer.

(D) phospholipid bilayer.

14. Substance x passes through a plasma membrane easily. What phrase best describes the probable nature of the substance?

(A) It is hydrophilic and nonpolar.

(C) It is hydrophilic and polar.

(B) It is hydrophobic and polar.

(D) It is hydrophobic and nonpolar.

15. The plasma membrane is soluble to

(A) lipids.   (B) proteins.   (C) acids.   (D) nucleic acids.

16. Movement of particles across a biological membrane is enhanced by

(A) large particle size.

(C) particle charge.

(B) lipid solubility.

(D) lipophobic properties.

17. Bile is manufactured in the

(A) duodenum.   (B) liver.   (C) gall bladder.   (D) pancreas.

18. Bile that is NOT immediately needed for digestion is stored and concentrated within the

(A) liver.   (B) gallbladder.   (C) pancreas.   (D) ileum.

19. Bile aids in the digestion of

(A) amino acids.   (B) fats.   (C) starches.   (D) carbohydrates.

20. In diseases of the gallbladder, which of the following nutrients is limited?

(A) starches   (B) proteins   (C) fats   (D) carbohydrates

21. Fat is broken down in the duodenum by _____ from the gallbladder.

(A) lipase   (B) amylase   (C) bile   (D) HCl

Proteins

Directions: Each question or incomplete statement below is followed by four suggested answers of completions. Select the best answer choice.

1. The substance basic to life is
   (A) carbohydrates.   (B) proteins.   (C) starches.   (D) fats.

2. Organic substances made up of several amino acids bound together are
   (A) carbohydrates.   (C) proteins.
   (B) fats.   (D) fatty acids.

3. Proteins are polymers of
   (A) hydrocarbons.   (C) heterocyclics.
   (B) amino acids.   (D) alcohols.

4. The basic building blocks of proteins are
   (A) polypeptides.   (C) amino acids.
   (B) glucose.   (D) None of the above

5. The number of different amino acids in proteins is
   (A) 20   (B) 26   (C) 50   (D) 92

6. A covalent bond between 2 amino acid molecules can be created by
   (A) inserting a water molecule between them.
   (B) removing a water molecule from them.
   (C) inserting a carbon atom between them.
   (D) removing a carbon atom from one of them.

7. The chemical bond that forms between the carboxyl (RCOOH) group of one amino acid and the amino (RC-NH$_2$) of another is a/an
   (A) peptide bond.   (C) ionic bond.
   (B) coordinate covalent.   (D) high energy bond.

8. In the small intestine, a digestive enzyme can break the peptide bond between two amino acids in a protein molecule by
   (A) removing a water molecule from them.
   (B) inserting a water molecule between them.
   (C) removing a carbon atom from one of them.
   (D) inserting a carbon atom between them.

9. _____ is the fibrous protein found in the stratum corneum that helps give the epidermis its protective properties.
   (A) Keratin   (B) Melanin   (C) Carotene   (D) Hemoglobin

10. Of the following, a human blood disease that has been definitely shown to be due to a hereditary factor or factors is
    (A) pernicious anemia.
    (C) polycythemia.
    (B) sickle cell anemia.
    (D) leukemia.

11. Sickle cell anemia is a genetic disorder that involves an inability of
    (A) erythrocytes to contain a sufficient amount of hemoglobin.
    (B) bone marrow to produce a sufficient number of erythrocytes.
    (C) bone marrow to produce erythrocytes of normal size.
    (D) erythrocytes to carry a sufficient load of oxygen.

12. Hemoglobin is a molecule composed principally of
    (A) ferritin.   (B) amino acids.   (C) iron.   (D) myosin.

13. Anemia is a condition resulting from abnormally low levels of
    (A) platelets.
    (C) leukocytes.
    (B) plasma proteins.
    (D) erythrocytes or hemoglobin.

14. Hemoglobin forms abnormal long chains in
    (A) pernicious anemia.
    (C) aplastic anemia.
    (B) iron deficiency anemia.
    (D) sickle cell anemia.

Enzymes

Directions: Each question or incomplete statement below is followed by four suggested answers of completions. Select the best answer choice.

1. Which of the following bodily substances is a catalyst?
    (A) bile
    (B) hemoglobin
    (C) enzyme
    (D) mucus

2. An organic catalyst that enhances the chemical reaction is called
    (A) a fat.
    (B) a lactic acid.
    (C) a polysaccharide.
    (D) an enzyme.

3. A protein substance that initiates and accelerates a chemical reaction is called a(n)
    (A) gene.
    (B) enzyme.
    (C) hormone.
    (D) base.

4. Enzyme molecules are all of the following EXCEPT
    (A) lipids.
    (B) proteins.
    (C) macromolecules.
    (D) biological catalysts.

5. Which is NOT a characteristic of enzymes?
    (A) They are proteins.
    (B) They catalize metabolics reactions.
    (C) They act on substances.
    (D) They are phospholipids.

6. Most human enzymes function best in the temperature range of
    (A) 5–15 degrees C.
    (B) 20–30 degrees C.
    (C) 35–40 degrees C.
    (D) 45–50 degrees C.

Questions 7-8 refer to the following diagram.

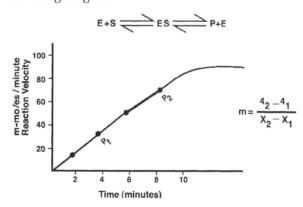

7. The linear distance between P_1 and P_2 on the above graph is the initial velocity of the reaction, as expressed by the slope (m) of the line (see above formula), which is
    (A) 60
    (B) 10
    (C) 40
    (D) 6

8. Choose the correct statement, relative to 10 minutes of reaction time.
    (A) Enzyme has been denatured.
    (B) All ES has been converted to P.
    (C) All enzyme E is occupied with substrate S.
    (D) All product has been converted to substrate.

9. Saliva contains mucus, water, and _____, which partially digests polysaccharides.
    (A) lipase   (B) amylase   (C) pepsin   (D) insulin

10. Protein digestion begins in the stomach by a group of proteolytic enzymes referred to as
     (A) lipase.   (B) sucrase.   (C) pepsin.   (D) amylase.

11. Of the following, an enzyme responsible for the digestion of proteins is
     (A) maltase.   (B) trypsin.   (C) ptyalin.   (D) steapsin.

12. The reaction of carbon dioxide and water leading to the formation of carbonic acid is catalyzed by
     (A) antigens on the surface of erythrocytes.   (C) enzymes in plasma.
     (B) antibodies within plasma.   (D) carbonic anhydrase within erythrocytes.

Nucleic Acids

SYNONYMS

Directions: In each of the sentences below, one word is in italics. Following each sentence are four words or phrases. For each sentence, select the word or phrase that best corresponds in meaning to the italicized word.

1. Production of complex molecules is accomplished by *replication*.
    (A) duplication   (B) synthesis   (C) fixation   (D) reproduction

2. The most important *mutation* is one occurring in the gametes.
    (A) fertilization   (B) meiosis   (C) deamination   (D) change

ANTONYMS

Directions: For each of the following test items, select the word that is opposite in meaning to the term printed in capital letters.

3. TERMINATE
    (A) withhold   (B) construe   (C) repel   (D) initiate

4. IMMUTABLE
    (A) erudite   (D) fantastic
    (B) abject   (E) aura
    (C) changeable

Directions: Each question or incomplete statement below is followed by four suggested answers of completions. Select the best answer choice.

5. Each nucleotide in a DNA molecule contains
    (A) a sugar.   (C) a phosphate group.
    (B) a nitrogen base.   (D) All of the above

6. A nucleotide is
    (A) phospholipid, sugar, and base.   (C) phosphate, protein, and base.
    (B) phosphate, sugar, and base.   (D) phospholipid, sugar, and protein.

7. Completion of the human genome project gave further proof that genes are
    (A) nucleic acid.   (C) protein.
    (B) phospholipids.   (D) protein and nucleic acid.

8. When completed, the number of DNA base pairs comprising the human genome is expected to be approximately
    (A) 10 thousand.   (B) 10 million.   (C) 3 billion.   (D) 3 trillion.

9. Data from the human genome project have shown that the total number of genes in humans is close to

   (A) 30   (B) 1,000,000   (C) 3,000   (D) 30,000

10. The Central Dogma of Information Transfer states that information is passed in what sequence?

    (A) RNA to proteins to DNA
    (B) DNA to RNA to proteins
    (C) Proteins to RNA to DNA
    (D) RNA to DNA to proteins

11. A segment of a DNA molecule transcribes a base sequence, AGAUAU, on an mRNA codon. The compatible sequence on the tRNA anticodon is

    (A) UUAGCG   (B) UCUAUA   (C) AAUAUA   (D) CGCAAA

12. Cellular proteins are synthesized in

    (A) ribosomes.   (B) mitochondria.   (C) lysosomes.   (D) golgi.

13. The biochemical technique that would be used to separate differently sized pieces of DNA created by digesting a sample of DNA with a restriction enzyme is

    (A) diapedesis.
    (B) liquid scintillation.
    (C) electrophoresis.
    (D) kinesthesia.

14. Small circular self-duplicating DNA molecules found in bacterial cells are

    (A) microbodies.
    (B) oligaproteins.
    (C) plasmids.
    (D) None of the above

15. Enzymes necessary for fragmenting genes to be cloned are

    (A) restriction enzymes.
    (B) splicing enzymes.
    (C) recombinases.
    (D) polymerases.

16. In the formation of recombinant DNA, the DNA from two different sources is spliced by an enzyme

    (A) polymerase.
    (B) DNA ligase.
    (C) recombinant dehydrogenase.
    (D) None of the above

17. Which of the following are requirements for cloning a gene?

    (A) An enzyme to fragment DNA
    (B) A vector (a bacterial plasmid or a virus)
    (C) A host cell or organism
    (D) All of the above

18. The polymerase chain reaction (PCR) technique is more efficient than cloning for copying large quantities of a gene because it is performed

    (A) without DNA polymerase.
    (B) in vitro.
    (C) without primers.
    (D) in vivo.

19. Hereditary determiners are found in

    (A) PKU.   (B) RNA.   (C) DNA.   (D) SMA.

Nutrition and Digestion

Directions: Each question or incomplete statement below is followed by four suggested answers of completions. Select the best answer choice.

1. Caloric needs are highest during
    (A) infancy.   (B) childhood.   (C) adulthood.   (D) middle age.

2. Which of the following foods is the most economical source of proteins?
    (A) Dried milk   (C) Meats
    (B) Green leafy vegetables   (D) Eggs

3. Diets in the United States are most often deficient in
    (A) iron and calcium.   (C) iodine and sodium.
    (B) calcium and potassium.   (D) phosphorous and iron.

4. Milk is not a "perfect food" because it lacks
    (A) iron.   (B) calcium.   (C) phosphorous.   (D) carbohydrates.

5. Amino acids that cannot be manufactured by the body are called
    (A) essential amino acids.   (C) basic amino acids.
    (B) synthetic amino acids.   (D) dependent amino acids.

6. The process by which the body changes food into substances that can be readily used by the body is
    (A) digestion.   (B) deglutition.   (C) micturition.   (D) absorption.

7. Digestion in humans is
    (A) extracellular.   (B) intracellular.   (C) vacuolar.   (D) intercellular.

8. A biochemical reaction, common to the digestion of carbohydrates, lipids, and proteins, is enzymatic
    (A) fermentation.   (B) deamination.   (C) glycogenolysis.   (D) hydrolysis.

9. In humans, the digestion of carbohydrates begins in the
    (A) stomach.   (B) small intestine.   (C) mouth.   (D) liver.

10. The greatest amount of energy would be produced by the burning of one gram of
    (A) fat.   (C) protein.
    (B) carbohydrate.   (D) ribonucleic acid.

11. Fats yield more calories per gram and oxidize slower than carbohydrates, proteins, and nucleic acids due to excess atoms of
    (A) hydrogen.   (B) oxygen.   (C) nitrogen.   (D) phosphorus.

Question 12 refers to the following compound.

$$H-\underset{\underset{\displaystyle H}{|}}{\overset{\overset{\displaystyle H}{|}}{C}}-\underset{\underset{\displaystyle NH_2}{|}}{\overset{\overset{\displaystyle H}{|}}{C}}-COOH$$

12. In human digestion, this compound is the end product of
    (A) fats.   (B) vitamins.   (C) proteins.   (D) carbohydrates.

13. Before amino acids can be metabolized to release energy, which of the following must occur?
    (A) fermentation   (B) hydrolysis   (C) deamination   (D) anabolism

14. The end product of protein metabolism is
    (A) amino acids.   (B) glucose.   (C) glycogen.   (D) fatty acids.

15. Urea formation is the human body's method of eliminating excess
    (A) carbon.   (B) hydrogen.   (C) nitrogen.   (D) phosphorus.

16. _____ are examples of substances that are NOT significantly reabsorbed by the kidneys and are therefore excreted.
    (A) Amino acids
    (B) Vitamins
    (C) Nitrogenous products of protein catabolism
    (D) Water and electrolytes

17. Urea is removed from the blood as it goes through the
    (A) bladder.   (B) pancreas.   (C) spleen.   (D) kidney.

18. In order to be absorbed, carbohydrates must be reduced to the form of
    (A) monosaccharides.   (C) polysaccharides.
    (B) disaccharides.   (D) oligosaccharides.

19. Carbohydrates are absorbed into the blood as
    (A) glycogen.   (B) amino acids.   (C) glucose.   (D) fatty acids.

20. Vitamins are important to the human diet because they are incorporated into
    (A) enzyme substitutes.   (C) co-enzymes.
    (B) ATP.   (D) inhibitors.

21. Diet pills on the market claim to reduce the amount of fat absorbed by 30 percent. The nurse explains to the young woman interested in taking them that they also reduce the amount of fat-soluble vitamins that are absorbed. These include
    (A) Vitamins B_1, B_6, and B_{12} (C) Vitamin C
    (B) Vitamins A, D, E, and K   (D) Calcium and iron

22. Water soluble vitamins include
    (A) vitamin A.   (B) vitamin C.   (C) vitamin D.   (D) vitamin K.

23. Vitamin _____ is water-soluble and readily absorbed from the intestine by diffusion.
    (A) A   (B) D   (C) E   (D) C

24. Another name for vitamin B_1 is
    (A) niacin.   (B) thiamin.   (C) riboflavin.   (D) pyridoxine.

25. The vitamin known as the "sunshine" vitamin is
    (A) vitamin E.   (B) vitamin B.   (C) vitamin K.   (D) vitamin D.

26. Vitamin C prevents
    (A) beriberi.   (B) rickets.   (C) pellagra.   (D) scurvy.

27. The vitamin that is necessary for coagulation of the blood is
    (A) vitamin K.   (B) vitamin C.   (C) vitamin A.   (D) vitamin E.

28. The vitamin that helps clotting of the blood is
    (A) C   (B) D   (C) E   (D) K

29. The vegetarian presents with anemia, fatigue, and loss of sensation in her hands and feet. The woman states that she does not eat any meat, chicken, or fish. The nurse, suspecting a vitamin B_{12} deficiency, ask if she includes the following in her diet:
    (A) Green, leafy vegetables   (C) Nuts, seeds, and dried fruits
    (B) Fresh fruits   (D) Eggs and milk

30. Of the following, vitamin B_{12} is most useful in combating
    (A) pernicious anemia.   (C) rickets.
    (B) night blindness.   (D) goiter.

31. Of the 92 naturally occurring elements, the number found in the human body is closer to
    (A) 50   (B) 10   (C) 25   (D) 75

32. Which of the following elements is the most abundant in the human body?
    (A) carbon   (B) potassium   (C) nitrogen   (D) oxygen

33. Iron is needed for
    (A) development of nervous tissue.   (C) growth of hair and nails.
    (B) formation of red blood cells.   (D) utilization of vitamins.

34. The largest portion of the iron supplied to the body by foods is used by the body for the
    (A) growth of hard bones and teeth. (C) development of respiratory enzymes.
    (B) manufacture of insulin.   (D) formation of hemoglobin.

35. The mineral that is necessary for the proper functioning of the thyroid gland is
    (A) sodium.   (B) iodine.   (C) calcium.   (D) iron.

36. Mr. Johnson, a 68 year old man with congestive heart failure, has been prescribed a low-sodium diet. In instructing him on appropriate food choices, which would the nurse counsel him *against* eating?
    (A) Spinach salad
    (B) Canned chicken noodle soup
    (C) Whole wheat bread
    (D) Apples

SYNONYMS

Directions: In the sentence below, one word is in italics. Following the sentence are four words or phrases. Select the word or phrase that best corresponds in meaning to the italicized word.

37. *Biotin* is widely distributed in organisms.
    (A) vitamin H   (B) vitamin D   (C) chyme   (D) plasmodia

ANTONYMS

Directions: For each of the following test items, select the word that is opposite in meaning to the term printed in capital letters.

38. SATIATED
    (A) satirical
    (B) centaur
    (C) gorgeous
    (D) delectable
    (E) hungry

Energy Storage and Usage

Directions: Each question or incomplete statement below is followed by four suggested answers of completions. Select the best answer choice.

1. The process in which carbon dioxide and water is combined under the influence of light in green plants is called
   (A) respiration.
   (B) fermentation.
   (C) assimilation.
   (D) photosynthesis.

2. The process responsible for the continuous removal of carbon dioxide from the atmosphere is
   (A) respiration.
   (B) metabolism.
   (C) oxidation.
   (D) photosynthesis.

3. The basic inorganic raw materials for photosynthesis are
   (A) water and oxygen.
   (B) water and carbon dioxide.
   (C) oxygen and carbon dioxide.
   (D) sugar and carbon dioxide.

4. In photosynthesis, the reactants CO_2 and H_2O, in the presence of sunlight and chlorophyll, combine chemically to produce glucose and O_2. The O_2 comes from
   (A) H_2O.
   (B) CO_2.
   (C) CO_2 and H_2O.
   (D) N_2.

5. In the equation below for photosynthesis, the oxygen comes $CO_2 + H_2O \rightarrow C_6H_{12}O_6 + O_2$
   (A) entirely from CO_2.
   (B) from a simple sugar molecule.
   (C) partially from CO_2 and H_2O.
   (D) entirely from H_2O.

6. Photosynthesis is a cellular process, which
   (A) is an exergonic reaction.
   (B) produces simple sugar and O_2.
   (C) is initiated by chemical energy.
   (D) produces CO_2 and H_2O.

7. The products of the light reaction of photosynthesis are
   (A) carbohydrate + CO_2.
   (B) $NADPH_2$ + ATP + O_2.
   (C) PGAL + CO_2 + H_2O.
   (D) starch + CO_2.

8. Which of the following is/are not a requirement for photosynthesis?
   (A) oxygen
   (B) carbon dioxide and water
   (C) sunlight
   (D) chlorophyll

9. The selection of algae as a possible additional food for humans is based primarily on their ability to carry on
   (A) fermentation.
   (B) digestion.
   (C) photosynthesis.
   (D) oxidation.

10. The most important of the greenhouse gases contributing to global warning and altering the marine carbon cycle is
    (A) SO_2
    (B) CO
    (C) NH_3
    (D) CO_2

11. In which one of the following ways does combustion differ from cellular respiration?
    (A) More heat is produced.   (C) It is less rapid.
    (B) More energy is wasted.   (D) It occurs at a higher temperature.

12. Identify the statement that is NOT true of cellular respiration.
    (A) It is a downhill process.
    (B) It occurs in both plant and animal cells.
    (C) It uses CO_2 and H_2O for reactants.
    (D) It is an exergonic process.

Questions 13 and 14 refer to the following diagram.

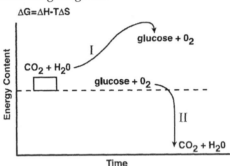

In cellular metabolism, the energy available for doing
biological work is called free energy (G). It is equal to the
molecular energy, enthalpy (H) minus the disorder,
entropy (S) times the absolute temperature (T).

13. Choose the correct statement describing reaction I.
    (A) The reaction is exergonic.
    (B) ΔG is negative.
    (C) Glucose and O_2 have less energy than CO_2 and H_2O.
    (D) The reaction is endergonic.

14. Choose the correct statement describing reaction II.
    (A) ΔG is positive.
    (B) CO_2 and H_2O have more energy than glucose and O_2.
    (C) The reaction is exergonic.
    (D) The reaction is endergonic.

15. The most efficient cellular respiratory process, in terms of energy-yield per molecule of glucose, is
    (A) aerobic respiration.   (C) fermentation.
    (B) anaerobic respiration.   (D) phosphorylation.

16. Aerobic cellular respiration is more important to sustaining life than anaerobic because it produces
    (A) more pyruvic acid.   (C) more energy.
    (B) more sugar.   (D) more lactic acid.

17. The process whereby muscle cells produce lactic acid is called
    (A) aerobic respiration.   (C) fermentation.
    (B) glycolysis.   (D) electron transport chain.

18. In the absence of oxygen, plants and microbes convert pyruvic acid into
    (A) alcohol and CO_2.   (C) CO_2 and H_2O.
    (B) lactic acid.   (D) amino acids.

19. During aerobic respiration, which one of the following substances is released?
    (A) 22 ATP   (B) 32 ATP   (C) 2 ATP   (D) 36 ATP

20. How many molecules of ATP are required to activate a molecule of glucose in glycolysis?
    (A) 2   (B) 6   (C) 3   (D) None

21. When a molecule of glucose in humans is degraded, the percent of its energy capable of generating ATP is nearest to
    (A) 100   (B) 50   (C) 25   (D) 80

22. Relaxed muscles produce more energy than is required for resting metabolism. That energy is stored in the form of
    (A) ATP.   (C) creatine.
    (B) ADP.   (D) creatine phosphate.

23. The chemical reaction that supplies immediate energy for muscular contractions can be summarized as
    (A) ATP → ADP + P.   (C) lactic acid → glycogen.
    (B) lactic acid → CO_2 + H_2O. (D) glycogen → ATP.

24. The first stage of aerobic cellular respiration is
    (A) electron transport chain.   (C) glycolysis.
    (B) Krebs cycle.   (D) light reaction.

25. Aerobic oxidation of glucose occurs in two major stages; these are
    (A) glycolysis and reduction.   (C) glycolysis and the Krebs cycle.
    (B) synthesis and the Krebs cycle. (D) degradation and hydrolysis.

26. The end products of the Krebs cycle are
    (A) carbon dioxide and water.   (C) lactic acid and pyruvic acid.
    (B) urea and bile pigments.   (D) ketones and acetones.

27. The Krebs cycle produces
    (A) H_2O and NADH.   (C) pyruvic acid and lactic acid.
    (B) CO_2 and H^+.   (D) amino acids.

28. As electrons from aerobic respiration are moved through the electron transport system (ETS), the final electron acceptor is
    (A) hydrogen.   (B) carbon dioxide.   (C) oxygen.   (D) water.

29. In cellular metabolism, glycolysis
    (A) requires O_2.
    (B) does not require O_2.
    (C) occurs only in animal cells.
    (D) produces $CO_2 + H_2O$.

30. Cellular energy production takes place within
    (A) rough endoplasmic reticulum.
    (B) smooth endoplasmic reticulum.
    (C) mitochondria.
    (D) Golgi apparatus.

31. With which organelle is the synthesis of ATP associated?
    (A) ribosome   (B) plastid   (C) mitochondrion   (D) lysosome

32. For the aerobic pathway, electron transport systems are located in the
    (A) cytoplasm.   (B) Golgi bodies.   (C) lysosomes.   (D) mitochondrion.

33. Glycolysis occurs in the
    (A) nucleus.
    (B) mitochondrion.
    (C) plasma membrane.
    (D) cytoplasm.

34. Glycogenesis occurs primarily in the
    (A) blood cells and spleen.
    (B) pancreas and gallbladder.
    (C) small intestines and stomach.
    (D) liver and muscles.

35. The concentration of glucose in blood cells is lower than the concentration of glucose in liver cells. During active transport, glucose moves from blood cells into the liver. Which organelle would you expect to find in large numbers in the liver?
    (A) Golgi bodies
    (B) Endoplasmic reticulum
    (C) Ribosomes
    (D) Mitochondria

Body Chemistry

Urine

Directions: Each question or incomplete statement below is followed by four suggested answers of completions. Select the best answer choice.

1. Electrolyte balance is maintained chiefly by the action of the
   - (A) bladder.
   - (B) kidney.
   - (C) islets of Langerhans.
   - (D) gonads.

2. Electrolyte balance is maintained primarily by the action of the
   - (A) testes.
   - (B) kidney.
   - (C) bladder.
   - (D) liver.

3. Choose the correct statement, relative to the distribution of sodium (Na^+) and potassium (K^+), on opposites of cell membranes.
   - (A) High Na^+ outside
   - (B) High K^+ outside
   - (C) Low Na^+ outside
   - (D) Low K^+ inside

4. The hormone most responsible for the renal regulation of sodium is
   - (A) thyroxine.
   - (B) insulin.
   - (C) glucagon.
   - (D) aldosterone.

5. The cation in the highest concentration within extracellular fluid that plays a major role in water distribution is
   - (A) sodium.
   - (B) potassium.
   - (C) calcium.
   - (D) magnesium.

6. Following an action potential, sodium is returned to the extracellular space through the process of
   - (A) diffusion.
   - (B) filtration.
   - (C) osmosis.
   - (D) active transport.

7. After rigorous exercise, the body is depleted of
   - (A) Na and H_2O.
   - (B) glucose and H_2O.
   - (C) H_2O and K.
   - (D) H_2O and colloids.

8. Water reabsorption in the collecting duct of the kidneys is controlled by _____ from the posterior pituitary.
   - (A) oxytocin
   - (B) ADH
   - (C) epinephrine
   - (D) aldosterone

9. Composition of urine normally includes
   - (A) creatinine, urea, and water.
   - (B) creatinine, ammonia, and sugar.
   - (C) nitrogen wastes, sugar, and hormones.
   - (D) nitrogen wastes, water, and pus cells.

10. Which of the following is abnormal for urine?
   - (A) Clear, amber liquid
   - (B) Nitrogenous waste products
   - (C) Slightly aromatic
   - (D) High specific gravity

11. The presence of protein in the urine is called
    (A) polyuria.   (B) anuria.   (C) albuminuria.   (D) hematuria.

12. There is concern that Mrs. Thompson is retaining fluid. How much urine output must this patient have per day to be considered within normal limits?
    (A) 1,000-1,500 cc   (B) 1,750-2,250 cc   (C) 5,000-10,000 cc   (D) 300-750 cc

13. Which reagent should be used in the urine test for diabetes?
    (A) Iodine
    (B) Nitric acid
    (C) Ammonia
    (D) Benedict's solution

Respiration

Directions: Each question or incomplete statement below is followed by four suggested answers of completions. Select the best answer choice.

1. Movement of respiratory gases is primarily dependent upon the presence of a(n) _____ gradient.
   (A) concentration
   (B) osmotic
   (C) partial pressure
   (D) temperature

2. Oxygen transported in blood is mainly
   (A) dissolved in plasma.
   (B) combined with hemoglobin.
   (C) CO_2.
   (D) carried as bicarbonate.

3. Oxygen unloading at the tissue level is accelerated by
   (A) elevated blood pressure.
   (B) lowered blood pressure.
   (C) elevated pH.
   (D) decreased pH.

4. CO_2 in blood is mainly
   (A) dissolved in plasma.
   (B) carried as bicarbonate.
   (C) dissolved in RBCs.
   (D) combined with hemoglobin.

5. Most of the carbon dioxide in the blood is carried in the
   (A) liquid portion.
   (B) leucocytes.
   (C) erythrocytes.
   (D) platelets.

6. Although most carbon dioxide is transported as bicarbonate, and some is dissolved in plasma, about 20% is carried
   (A) by leukocytes.
   (B) by plasma proteins.
   (C) as carbaminohemoglobin.
   (D) on the surface of erythrocytes.

7. The exchange of carbon dioxide and oxygen in the lungs occurs in the
   (A) venules.   (B) alveoli.   (C) bronchi.   (D) bronchioles.

8. The deadly property of carbon monoxide, if inhaled, is due to its
   - (A) high affinity for O_2.
   - (B) low affinity for hemoglobin.
   - (C) high affinity for hemoglobin.
   - (D) conversion to cyanide gas.

9. Hyperventilation resulting from hysteria may cause
   - (A) respiratory acidosis.
   - (B) respiratory alkalosis.
   - (C) metabolic acidosis.
   - (D) metabolic alkalosis.

Blood

Directions: Each question or incomplete statement below is followed by four suggested answers of completions. Select the best answer choice.

1. Normal pH of blood is
   - (A) 6
   - (B) 7.4
   - (C) 8
   - (D) 1.0

2. The most rapid mechanism of pH adjustment involves
   - (A) buffers.
   - (B) the respiratory system.
   - (C) the kidneys.
   - (D) the liver.

3. Acid-base regulation is most powerful and most complete by
   - (A) buffers.
   - (B) respiratory mechanisms.
   - (C) renal mechanisms.
   - (D) urinary bladder activity.

4. The liquid portion of the blood is called
   - (A) serum.
   - (B) gamma globulin.
   - (C) plasma.
   - (D) lymph.

5. The major difference between plasma and blood is
   - (A) cellular content.
   - (B) acid-base balance.
   - (C) anion-cation placement.
   - (D) solute-solvent concentrations.

6. The major component of plasma is
   - (A) ions.
   - (B) proteins.
   - (C) water.
   - (D) gases.

7. The hormone that regulates blood composition and blood volume by acting on the kidney is
   - (A) antidiuretic (ADH).
   - (B) aldosterone.
   - (C) parathormone.
   - (D) oxytocin.

8. The force of the blood exerted against the wall of the blood vessel is called
   - (A) pulse deficit.
   - (B) apical pulse.
   - (C) blood pressure.
   - (D) pulse pressure.

9. All of the following mechanisms affect the amount of glucose in the blood EXCEPT
   - (A) adrenaline secretion.
   - (B) insulin secretion.
   - (C) level of oxygen intake.
   - (D) level of physical activity.

10. The hormone from the pancreas that is responsible for elevating blood glucose levels between meals is
   (A) insulin.   (B) glucagon.   (C) somatostatin.   (D) epinephrine.

11. Pancreatic beta cells produce the hormone _____, which lowers blood glucose levels by promoting glucose uptake and utilization.
   (A) glucagon   (B) insulin   (C) epinephrine   (D) renin

12. Insulin is produced in the
   (A) pituitary gland.   (C) pancreas.
   (B) thymus.   (D) pineal gland.

13. Insulin is released by the beta cells of the pancreas in response to
   (A) low blood glucose levels.   (C) high prolactin levels.
   (B) high blood glucose levels.   (D) high growth hormone levels.

14. Diabetes mellitus initially results from
   (A) oversecretion of pancreatin.   (C) excessive intake of sugar.
   (B) undersecretion of insulin.   (D) inadequate intake of fats.

15. Type I diabetes mellitus is characterized by
   (A) high insulin levels.   (C) low blood glucose levels.
   (B) ineffective insulin receptors.   (D) decreased insulin production.

16. Mrs. Farinella has been newly diagnosed with insulin dependent diabetes mellitus. You must teach her the symptoms of hypoglycemia. Which would be the best description?
   (A) Increased heart rate, hunger, sweating, tremors, and confusion.
   (B) Dry mouth, nausea, dizziness, and tremors.
   (C) Hunger, lethargy, difficulty breathing, and increased urine output.
   (D) Sweating, decreased heart rate, lethargy, and nausea.

17. The patient, who is 62-years-old, overweight, and has a family history of diabetes presents to the nurse for her first follow-up visit after her diet and exercise plan has been put in place. Without checking the chart, the nurse knows that this patient most likely has
   (A) Type I diabetes.   (C) gestational diabetes.
   (B) Type II diabetes.   (D) impaired glucose tolerance.

18. The nurse is caring for a patient who has been newly diagnosed as diabetic. The care plan includes careful monitoring of intake and output. The nurse will expect to find which of the following, which is a typical presentation of diabetes?
   (A) anuria   (B) polyuria   (C) hematuria   (D) oliguria

19. The clinic nurse is evaluating a man wearing a diabetic medic alert band, who appears confused, with hot, dry, flushed skin. His repirations are deep and fast, and he says he is nauseated. His breath smells fruity. The nurse recognizes that he needs immediate care because he has the following diabetes-related condition:
   (A) ketoacidosis   (B) hypoglycemia   (C) neuropathy   (D) retinopathy

20. Uncontrolled diabetes mellitus results in
    (A) respiratory acidosis.
    (C) metabolic acidosis.
    (B) respiratory alkalosis.
    (D) metabolic alkalosis.

Health and Medicine

Directions: Each question or incomplete statement below is followed by four suggested answers of completions. Select the best answer choice.

1. A *histamine* is released from the tissues when the cells are injured.
   (A) a histone   (B) an amine   (C) a stimulant   (D) an isoenzyme

2. The stimulant in coffee is
   (A) tannic acid.   (B) theobromine.   (C) theophylline.   (D) caffeine.

3. Skin color varies with the amount of
   (A) melanin.   (B) matrix.   (C) hair.   (D) keratin.

4. The thyroid gland cannot function properly without
   (A) chloride.   (B) iodine.   (C) phosphorus.   (D) iron.

5. A direct physiological effect of radiation on human tissues is
   (A) impairment of cellular metabolism.
   (C) formation of scar tissue.
   (B) proliferation of white blood cells.
   (D) reduction of body fluids.

6. Two weeks after receiving chemotherapy, Mrs. Constant develops sores in her mouth, has hair loss, and complains of being extremely tired. She asks what is happening to her. Your best answer is
   (A) "The chemotherapy is designed to attach rapidly multiplying cells but doesn't distinguish between cancer cells and normal cells.
   (B) "Chemotherapy is very toxic, and she was told she would experience side effects before the therapy was started."
   (C) "These are normal side effects, and they will go away after awhile."
   (D) "Give her a booklet she can read, and tell her to ask her doctor if she has questions."

7. In patients with cystic fibrosis, the accumulation of mucus around the membranes of cells in the lungs, liver, and pancreas has been shown to be caused by the blockage of ionic channels for
   (A) Ca^{2+} (B) Cl^- (C) K^+ (D) Na^+

8. Phenylketonuria is a genetic disorder that involves an inability of
   (A) blood to clot properly.
   (B) one amino acid to be converted to another.
   (C) blood cells to carry a sufficient load of oxygen.
   (D) lung alveoli to stay open.

9. Slow muscle fibers are particularly resistant to fatigue, a property due in part to the presence of _____, a protein that reversibly binds oxygen.
   (A) hemoglobin   (B) myoglobin   (C) ATP   (D) ADP

Reading Comprehension for Science Topics

Directions: Carefully read the following passage and then answer the accompanying questions, basing your answers on what is stated or implied in the passage. Select the best answer choice for each question. There is only one best answer for each question.

DNA

With the recent use of DNA (deoxyribonucleic acid) as a means of providing evidence in a number of world-renowned criminal cases, the general public views this carrier of genetic information as a modern day scientific discovery. However, a glimpse at the history of DNA will prove the notion of a "modern day miracle" quite to the contrary.

To investigate the discovery of DNA, one would have to research the laboratory and work of the Swiss biochemist, Johan Friedrich Miescher, back in 1868. Miescher had been involved in the study of the cell nucleus, the round control center that contains the chromosomes as well as other elements. He believed that cells were made of protein and attempted to break down this protein with a digestive enzyme. As Miescher continued this investigation, he was perplexed by the fact that the enzyme would break down the cell but not the nucleus. He then launched an investigation of the substance that comprised the cell. As he analyzed it, he saw that it contained large amounts of a strange material that was very unlike protein. Miescher chose to call this substance nuclein. He had no idea of its significance, nor did he recognize that he had discovered what came to be known in later years as nucleic acid. Nucleic acid is the chemical family to which DNA belongs.

In 1944 a team of scientists from the Rockefeller Institute proved for the first time that DNA was the carrier of hereditary information. Oswald T. Avery, Colin M. MacLeod, and Maclyn McCarty accomplished this by extracting some of the DNA in pure form from a bacterium and substituting it for a defective gene in another related bacterium.

Some ten years later, the intricate molecular structure of DNA was described by Harvard biochemist James D. Watson and physicist Francis Crick of Great Britain. However, prior to this, scientist Rosalind Franklin discovered that the DNA molecule was a strand of molecules in a spiral form. Dr. Franklin demonstrated that the spiral was so large that it was most likely formed by two spirals. Ultimately, Franklin determined that the structure of DNA is similar to the handrails and steps of a spiral staircase.

Equipped with the work and findings of Rosalind Franklin and others, Watson and Crick were able to construct a model of a DNA molecule. This model depicted the sides or "handrails" of the DNA molecules as being made up of two twisted strands of sugar and phosphate molecules. The "stairs" that hold the two sugar phosphate strands apart are made up of molecules called nitrogen bases.

All of this data supports the fact that DNA is by no means a "new discovery"; however, what is the significance of it at all? Why is DNA important to you? The answer is that all of the characteristics you possess are affected by the DNA in your cells. It controls the color of your eyes, the color of your hair, and whether or not you have a tolerance for dairy products. These characteristics are known as traits. The way your traits appear depends on the kinds of proteins your cells make. DNA stores the blueprints for making the proteins. Your DNA is uniquely different from that of anyone else on earth, and you are identifiable by these proteins.

1. It could be concluded that
    (A) Watson and Crick discovered DNA.
    (B) the strands of DNA take the form of a double hexagon.
    (C) DNA is as unique to individuals as a fingerprint.
    (D) Miescher's analysis of nuclein resulted directly in the discovery of DNA.

2. From this passage, it may be deduced that enzymes are
    (A) unstable.   **(B)** ineffective.   **(C)** catalysts.   **(D)** solutions.

3. It may be inferred that an individual's DNA determines
    (A) whether or not he or she can digest milk.
    (B) whether or not he or she is immune to the common cold.
    (C) an individual's choice of residential location.
    (D) a person's inclination toward dishonesty.

4. The types of protein produced by a cell are controlled by the DNA contained in its
    (A) nitrogen bases.   **(C)** cell wall.
    (B) nucleus.   **(D)** sugar phosphate bands.

5. It is implied that proteins are
    (A) the control center of the cell.   **(C)** the storage center of the cell.
    (B) the blueprint of the cell.   **(D)** the building blocks of a cell.

6. A reference to the "spiral staircase" constitutes a description of
    (A) the molecular structure of proteins.
    (B) the molecular structure of digestive enzymes.
    (C) the molecular structure of RNA.
    (D) the molecular structure of DNA.

7. Digestive enzymes are effective in breaking up
    (A) nuclein.   **(C)** all chemical compounds.
    (B) DNA.   **(D)** protein.

8. The word "nucleus" refers to
    (A) the round control center of the cell.   **(C)** a strand of molecules.
    (B) the walls of the cell.   **(D)** a helix.

9. The scientific disciplines used in determining the structure of the DNA molecule include biology, chemistry, and
    (A) genealogy.   **(B)** serology.   **(C)** physics.   **(D)** embryology.

10. As described in this passage, "model" means
    (A) an exhibitor of fashion.
    (B) a physical form representing a concept.
    (C) a miniature version of an existing object.
    (D) a person on whom an artist bases his or her rendition.

Directions: Carefully read the following paragraphs and then answer the accompanying questions, basing your answer on what is stated or implied in the paragraphs. Select the best answer choice for each question. There is only one best answer for each question.

Is It Your Fault If You're Fat?

A

Why are more Americans overweight and developing diabetes? Is it fast food? No regular meals and precious little exercise? Our love affair with the TV and computer? The wrong advice about what to eat?

B

Very likely it's all these things—combined with something about the genetic make-up in many of us. Genes may program some to feel hungry when they aren't and others to be less able to tell when they are full.

1. The generalization could be made that obesity is caused
   (A) by eating fast foods.
   (B) lack of exercise.
   (C) lack of education as to which foods to eat.
   (D) lifestyle and genetic make-up.

2. People are programmed to feel hungry (when they actually are not) by
   (A) therapists.
   (B) genes.
   (C) nutritional counseling.
   (D) dining trends.

C

Some extreme obesity in children is caused by an identifiable, single gene defect. Obesity is no more their "fault" than developing cystic fibrosis is the "fault" of a child who has the CF gene. Admittedly, perhaps only 5 percent of obesity is purely genetic. But research suggests that multiple genes control appetite and metabolism, and defects in one or more may make someone more prone to being overweight. Fat cells, particularly in the abdomen, in turn release substances that can make people more prone to insulin resistance, which leads to Type 2 diabetes.

D

Some people are genetically blessed and never gain much weight. Those with gene defects must expend huge effort to overcome messages their bodies are sending their brains to eat more.

3. People who are more prone to insulin resistance are
   (A) candidates for stomach problems.
   (B) those with no weight problem.
   (C) those who are likely to develop Type 2 diabetes.
   (D) in danger of becoming obese.

4. When great effort is put forth, people with gene defects can
    (A)   overcome urges to eat more.   (C)   live a normal life.
    (B)   overcome diabetes.   (D)   be overweight.

5. The author's purpose in this passage is to
    (A)   inform.   (B)   persuade.   (C)   entertain.   (D)   analyze.

E

For example, research at Rockefeller University and elsewhere suggests that people who lack leptin or lack receptors to make their cells sensitive to leptin have uncontrolled hunger, overeat, and become extremely obese. The melanocortin pathway in the brain has recently been identified by scientists at Beth Israel Deaconess Medical Center and elsewhere as another target influencing both obesity and anorexia.

F

In addition, some researchers have shown that the absence of a peptide called MSH, which suppresses eating, leads to obesity. Research at Joslin Diabetes Center in Boston and elsewhere suggests that another peptide, MCH, which stimulates eating, may also play a role.

6. Researchers have found that
    (A)   an overabundance of leptin causes obesity.
    (B)   the absence of a certain substance in the body results in uncontrolled eating.
    (C)   MHS is a peptide.
    (D)   overeating is a habit.

7. Obesity may be defined as
    (A)   fat.   (C)   a genetic condition.
    (B)   a state of being significantly overweight.   (D)   a disease related to overeating.

G

Ghrelin, a stomach hormone that signals hunger, is another potential target. Interestingly, extremely obese individuals who undergo stomach bypass surgery (stomach stapling) may be less inclined to eat afterwards because food no longer passes through the section of the stomach that produces ghrelin.

H

Some overeating may be triggered by stress, boredom, or depression. For example, food smells may stimulate production of certain peptides that make one want to eat—even if not hungry. Behavior modification may be needed to combat these stimulants of weight gain.

I

As our genes haven't changed in the last 20–30 years, societal influences are still the major culprit for growing obesity. We are more sedentary. Super-sizing meals makes it harder not to eat for those trying mightily to ignore the errant signals their bodies are sending. Until we can identify who has what gene defects, and the medications are developed to treat them, we must remember that it is much easier to prevent weight gain than to lose weight once gained. Your body adjusts quickly to these extra calories.

J

Life as overweight adults often has its roots in life as a child. For the moment, the best approach to obesity, and the type 2 diabetes it causes, is prevention—in ourselves and our children. As nationwide studies show, even modest weight loss—15 pounds—and 30 minutes of daily exercise, are the best ways to prevent diabetes in those most likely to develop it.

—Copyright © Joslin Diabetes Center. Reprinted with permission from *Time*, "Is It Your Fault If You're Fat" by Eleftheria Marasthos-Flier, M.D. and Jeffrey S. Flier, M.D. *Time* Magazine Special Advertising Section, November 4, 2002.

8. The main idea of this passage is
    (A) overeating can be controlled with diet and exercise.
    (B) diabetes is related to obesity.
    (C) current research reveals a number of contributing factors to obesity.
    (D) there are cures for obesity.

9. From this passage one can draw the conclusion that
    (A) being overweight is unhealthy.
    (B) children, too, can have weight problems.
    (C) being fat may not necessarily be the fault of the obese individual.
    (D) obesity can be attributed solely to genetics.

Directions: Carefully read the following passage and then answer the accompanying questions, basing your answers on what is stated or implied in the passage. Select the best answer for each question. There is only one best answer for each question.

Body Fluids

Every cell in the body is bathed in water. The substances dissolved in the water provide the immediate environment for the cells' existence, that is, for their respiration, digestion, excretion, and reproduction. This water—70% of our body weight and originally from our food and drink—is carried in the blood and is distributed in three places in the body. In terms of body weight:

> 5% remains in blood
> 15% goes to tissue spaces
> 50% goes inside the cell

Of course, more water is needed in the cell than elsewhere in the body because there are more solutes to be dissolved and because ionization must take place for anabolism and catabolism.

All of the substances inside the cell and outside the cell in the tissue spaces were at one time a part of the blood. The blood is the transportation system of the body and, therefore, is the recipient of all substances. Substances are "unloaded" by pushing out those that can filter through the tiny openings of the semipermeable membranes of the capillaries into the tissue spaces. These substances, along with water, are then sucked into the semipermeable membranes of the cells. At the same time, the cell has produced waste substances; it now pushes those out into the tissue fluid so that they can be pushed into the capillaries and thereby excreted by the kidneys, skin, respiratory system, etc.

The pushing and sucking forces are regulated by concentrations. The more concentrated a substance is, the more force it has. What determines whether that force will push or suck is whether the substance is primarily water (solvent) or particles dissolved in water (solute). If water is the primary component of a substance, it will be sucked; that is, it will pass from a lesser to a greater concentration. This is called osmosis. The substance that is sucking in the water has the greater force or pull, called osmotic pressure, because it has the greater concentration. If solutes are the primary component, they can be pushed only by diffusion from a greater to a lesser concentration. The sucking and pushing processes go on simultaneously, and that is why we identify this mechanism as being dynamic. The "equilibrium" of body fluids is explained in terms of a balance of forces. This dynamic factor is controlled by three things:

1. How much of the substance is present (solute/solvent)

2. What kind of substance is present (electrolytes/non-electrolytes)

3. The placement and distribution of the substance (cell/tissue space/blood)

The concentration of a solution is determined by the relationship between the solutes and water. "Solutes" may be electrolytes or non-electrolytes. Non-electrolytes do not ionize and, therefore, they affect the concentration of a solution and its diffusion processes, but they do not affect the osmolarity of a solution. Osmotic pressure is determined by "tonic" relationships. "Tonic" refers to the comparison of the number of specific ions per unit volume in two given solutions—iso being "the same as," hypo being "less than," and hyper being "more than."

Given two solutions of the same solute, the more concentrated solution contains more solute particles, has a higher potential osmotic pressure, and is hypertonic as compared with the less concentrated solution. Given two solutions of different solutes, the solution containing more particles has the greater concentration, but there are no "tonic" relationships.

1. Why is an adequate intake of water essential to life?
   (A) Solutes will dissolve only in water.
   (B) Water is the medium for exchange of solutes.
   (C) Osmosis will take place only in water.
   (D) Water is necessary for metabolism.

2. The blood is described as "the transportation system of the body" because
   (A) blood has the capacity for osmosis and diffusion.
   (B) all nutrients and wastes are received by the blood.
   (C) blood has a higher concentration than any other body fluid.
   (D) all metabolic processes are controlled by the blood.

3. The concentration of a solution is directly determined by
   (A) placement of solutes.
   (B) distribution of water.
   (C) percentage of solute to solvent.
   (D) percentage of electrolytes to non-electrolytes.

4. The largest percentage of H_2O in the body is found
   (A) intracellularly. (B) extracelluarly. (C) intravascularly. (D) interstitially.

5. In the process of osmosis (sucking), the primary factor is
   (A) water. (B) solute. (C) ionization. (D) force.

6. If blood cells were placed in a hypertonic solution of salt, which one of the following blood cell reactions would take place?
   (A) Swelling (B) Shrinkage (C) Destruction (D) No change

7. The primary factor in diffusion (pushing) is
   (A) water. (B) solute. (C) ionization. (D) force.

8. If there are two solutions of sodium chloride, which of the following factors will determine which solution has the highest potential for osmosis?
   (A) Isotonicity (B) Hypotonicity (C) Hypertonicity (D) All of the above

Directions: Carefully read the following paragraphs and then answer the accompanying questions, basing your answer on what is stated or implied in the paragraphs. Select the best answer choice for each question. There is only one best answer for each question.

Blood Donors Available—No Thanks! I'll Do It Myself!

A

With the advent of AIDS and other communicable diseases, people are reluctant to submit to blood transfusions as a measure of medical treatment, even in emergency situations. Many individuals are opting to store their own blood in case it is ever needed.

B

New blood cells are constantly being produced in the body. This is the reason that lost blood in a healthy person is replaced quickly. This rapid production of blood cells also enables a person to donate to others who might need blood. When blood is taken from one person and given to another, the procedure is called a homologous blood transfer. When a person's own blood is used for transfusion, having been stored in anticipation of surgery, the procedure is known as autologous blood transfer. This blood is collected prior to surgery.

C

Though the concept of being transfused with one's own blood is very comforting, there are many factors that discourage the use of this technology. First is the expense of storing the blood. It has been estimated that the cost of safely storing the blood comes to approximately $200 per year. Another drawback is that a patient may be thousands of miles away from his or her blood supply or blood bank when the need for blood arises.

1. The accelerated interest in autologous transfusion is due to
   (A) a lack of blood donors.
   (B) an attempt to stimulate rapid production of new blood cells.
   (C) the time-saving element in the event of emergency surgery.
   (D) a rise in fear of communicable diseases.

D

In an attempt to satisfy patients' requests to be transfused with their own blood, regardless of the medical emergency, doctors have developed two forms of autologous blood transfusion. One form utilizes suction devices to collect blood lost during surgery. After this blood has been cleansed, it may be put back into the body. Blood lost during surgery can also be collected with sponges. The sponges are then squeezed out into a container of saline solution, a kind of salt solution. This blood is processed within 15 minutes and is again introduced into the patient's circulation. Doctors believe that by using both methods they can retrieve up to 90 percent of blood that would otherwise be lost.

2. From this passage it may be inferred that
   (A) saline is a cleansing or sterilizing agent.
   (B) the storing of one's own blood supply is affordable for all.
   (C) homologous transfusion requires blood typing and matching.
   (D) the option of autologous transfusion is not feasible in instances of elective surgery.

E

Despite these efforts to make autologous blood transfer feasible for most patients, there still remain those situations that make it virtually impossible to collect blood. For instance, in the case of an auto accident victim, much blood is often lost before medical treatment is obtainable. To help in an incident such as this, doctors are researching ways to develop artificial hemoglobin that would temporarily transport oxygen and carbon dioxide throughout the body. Another solution being explored is the reproduction of a hormone that causes the body to produce blood cells much more rapidly than it would normally. This would mean that the body could replace most of its own blood and therefore reduce the need for transfusion. Even though researchers are doing extensive work to develop new techniques to protect people from blood tainted by disease, reserving one's own blood for future use seems to be the safest method of transfusion for now.

3. This passage supports the concept that
   (A) autologous transfer is a practical and easily accessible alternative to blood transfusion.
   (B) any type of blood transfusion places a patient in a high-risk health situation.
   (C) hemoglobin carries oxygen and carbon dioxide throughout the body.
   (D) lost blood, even in a healthy person, is replaced only after a long period of time.

4. A fact expressed in this passage is that
   (A) artificial hemoglobin could permanently supply the body with oxygen and carbon dioxide.
   (B) suction devices and sponges are two surgical implements used in the collection of blood lost during surgery.
   (C) laser surgery is being used more frequently in an effort to minimize blood loss during surgery.
   (D) autologous transfusion, in the event of surgery, requires that the blood always be collected prior to surgery.

5. This article suggests that
   (A) homologous transfusions are on the decline.
   (B) the effort by researchers to protect patients from blood contaminated by disease cannot guarantee safe blood transfusion.
   (C) autologous transfusion is impractical.
   (D) surgical patients who are transfused autonomously are assured a more rapid recovery.

Answers for Practice Questions

Mathematics and Measurement

1.	A	16.	C	31.	D	46.	B
2.	B	17.	A	32.	A	47.	B
3.	A	18.	B	33.	C	48.	D
4.	B	19.	D	34.	B	49.	C
5.	A	20.	D	35.	B	50.	D
6.	D	21.	C	36.	C	51.	C
7.	B	22.	C	37.	A	52.	C
8.	C	23.	A	38.	C	53.	B
9.	D	24.	D	39.	A	54.	D
10.	A	25.	D	40.	A	55.	A
11.	A	26.	C	41.	B	56.	C
12.	C	27.	C	42.	A	57.	A
13.	D	28.	B	43.	B	58.	D
14.	B	29.	C	44.	C	59.	D
15.	B	30.	C	45.	A	60.	B

Matter and the Periodic Table

1.	B	14.	B	27.	C	40.	B
2.	A	15.	C	28.	C	41.	B
3.	D	16.	A	29.	B	42.	D
4.	A	17.	A	30.	B	43.	D
5.	C	18.	B	31.	C	44.	C
6.	A	19.	A	32.	D	45.	D
7.	A	20.	B	33.	A	46.	B
8.	B	21.	A	34.	B	47.	A
9.	B	22.	B	35.	C	48.	C
10.	B	23.	A	36.	C	49.	A
11.	A	24.	B	37.	D	50.	C
12.	B	25.	D	38.	D	51.	D
13.	D	26.	A	39.	B		

Bonding, Structure, and Reactions

1.	C	13.	A	25.	D	37.	C
2.	C	14.	D	26.	B	38.	C
3.	B	15.	A	27.	C	39.	C
4.	A	16.	B	28.	D	40.	C
5.	B	17.	D	29.	D	41.	B
6.	C	18.	C	30.	B	42.	A
7.	C	19.	B	31.	B	43.	A
8.	D	20.	A	32.	D	44.	A
9.	D	21.	A	33.	B	45.	D
10.	C	22.	B	34.	A	46.	B
11.	C	23.	A	35.	D	47.	C
12.	D	24.	D	36.	C	48.	A

Gases, Boiling, and Specific Heat

1.	A	8.	A	15.	B	22.	A
2.	B	9.	D	16.	B	23.	D
3.	C	10.	D	17.	D	24.	D
4.	B	11.	B	18.	A	25.	A
5.	B	12.	A	19.	D	26.	A
6.	B	13.	B	20.	D		
7.	D	14.	A	21.	C		

Solutions

1.	C	8.	C	15.	B	22.	B
2.	B	9.	C	16.	C	23.	D
3.	C	10.	C	17.	C	24.	B
4.	D	11.	B	18.	A	25.	C
5.	B	12.	B	19.	D	26.	C
6.	B	13.	D	20.	C	27.	B
7.	D	14.	B	21.	D	28.	A

Acids, Bases, and Salts

1.	A	5.	D	9.	C	13.	A
2.	D	6.	D	10.	C	14.	D
3.	B	7.	C	11.	A	15.	B
4.	B	8.	C	12.	C		

Nuclear Chemistry

1.	C	4.	A	7.	B
2.	D	5.	D	8.	C
3.	C	6.	A	9.	A

Organic Chemistry

1.	D	5.	C	9.	B	13.	D
2.	B	6.	C	10.	D	14.	A
3.	A	7.	D	11.	D		
4.	D	8.	D	12.	C		

Carbohydrates

1.	D	4.	A	7.	D	10.	D
2.	C	5.	B	8.	C	11.	B
3.	B	6.	B	9.	D	12.	B

Lipids

1.	D	7.	D	13.	B	19.	B
2.	B	8.	C	14.	D	20.	C
3.	B	9.	C	15.	A	21.	C
4.	D	10.	C	16.	B		
5.	C	11.	A	17.	B		
6.	D	12.	B	18.	B		

Proteins

1.	B	5.	A	9.	A	13.	D
2.	C	6.	B	10.	B	14.	D
3.	B	7.	A	11.	D		
4.	C	8.	B	12.	B		

Enzymes

1.	C	4.	A	7.	B	10.	C
2.	D	5.	D	8.	C	11.	B
3.	B	6.	C	9.	B	12.	D

Nucleic Acids

1.	A	7.	A	13.	C	19.	C
2.	D	8.	C	14.	C		
3.	D	9.	D	15.	A		
4.	C	10.	B	16.	B		
5.	D	11.	B	17.	D		
6.	B	12.	A	18.	B		

Nutrition and Digestion

1.	A	11.	A	21.	B	31.	C
2.	A	12.	C	22.	B	32.	D
3.	A	13.	C	23.	D	33.	B
4.	A	14.	A	24.	B	34.	D
5.	A	15.	C	25.	D	35.	B
6.	A	16.	C	26.	D	36.	B
7.	A	17.	D	27.	A	37.	A
8.	D	18.	A	28.	D	38.	E
9.	C	19.	C	29.	D		
10.	A	20.	C	30.	A		

Energy Storage and Usage

1.	D	10.	D	19.	D	28.	C
2.	D	11.	D	20.	A	29.	B
3.	B	12.	C	21.	B	30.	C
4.	A	13.	D	22.	D	31.	C
5.	D	14.	C	23.	A	32.	D
6.	B	15.	A	24.	C	33.	D
7.	B	16.	C	25.	C	34.	D
8.	A	17.	C	26.	A	35.	D
9.	C	18.	A	27.	B		

Body Chemistry: Urine

1.	B	5.	A	9.	A	13.	D
2.	B	6.	D	10.	D		
3.	A	7.	A	11.	C		
4.	D	8.	B	12.	A		

Body Chemistry: Respiration

1.	C	4.	B	7.	B
2.	B	5.	A	8.	C
3.	D	6.	C	9.	B

Body Chemistry: Blood

1.	B	6.	C	11.	B	16.	A
2.	A	7.	B	12.	C	17.	B
3.	C	8.	C	13.	B	18.	B
4.	C	9.	C	14.	B	19.	A
5.	A	10.	B	15.	D	20.	C

Body Chemistry: Health and Medicine

1.	B	4.	B	7.	B
2.	D	5.	A	8.	B
3.	A	6.	A	9.	B

Reading Comprehension: DNA

1.	C	4.	B	7.	D	10.	B
2.	C	5.	D	8.	A		
3.	A	6.	D	9.	C		

Reading Comprehension: Is It Your Fault If You're Fat?

1.	D	4.	A	7.	B
2.	B	5.	A	8.	C
3.	C	6.	B	9.	C

Reading Comprehension: Body Fluids

1.	D	3.	C	5.	A	7.	B
2.	B	4.	A	6.	B	8.	C

Reading Comprehension: Blood Donors Available – No Thanks I'll Do It Myself

1.	D	3.	C	5.	B
2.	A	4.	B		

In-Depth Solutions for Mathematics and Measurement

1. **(A)** Area = length × width
   Big ▭ Area = 15 × 6 = 90
   Little ▭ Area = 9 × 4 = 36
   $90 - 36 = \boxed{54}$

2. **(B)** $40 \ \cancel{\text{inches}} \left(\dfrac{1 \text{ foot}}{12 \ \cancel{\text{inches}}}\right) = 3.\overline{3} \text{ feet}$ $20 \ \cancel{\text{inches}} \left(\dfrac{1 \text{ foot}}{12 \ \cancel{\text{inches}}}\right) = 1.\overline{6} \text{ feet}$
   Area = length × width = $3.\overline{3}$ feet × $1.\overline{6}$ feet = $5.\overline{5}$ feet ≈ $\boxed{5.5 \text{ feet}}$

3. **(A)** $25 \ \cancel{\text{feet}} \left(\dfrac{1 \text{ yard}}{3 \ \cancel{\text{feet}}}\right) = 8.\overline{3} \text{ yards}$ $36 \ \cancel{\text{feet}} \left(\dfrac{1 \text{ yard}}{3 \ \cancel{\text{feet}}}\right) = 12 \text{ yards}$
   Area = length × width = $8.\overline{3}$ yards × 12 yards = $\boxed{100 \text{ yd}^2}$

4. **(B)** Area = length × width
   $63 \text{ ft}^2 = 9 \text{ feet} \times \text{width}$
   $\text{width} = \dfrac{63 \text{ ft}^2}{9 \text{ ft}} = \boxed{7 \text{ ft}}$

5. **(A)** Area = length × width
   square length = square width
   Area = side^2
   $\text{side} = \sqrt{\text{area}} = \sqrt{144 \text{ m}^2} = \boxed{12 \text{ m}}$

6. **(D)** $8 \ \cancel{\text{inches}} \left(\dfrac{1 \text{ foot}}{12 \ \cancel{\text{inches}}}\right) = 0.\overline{6} \text{ feet}$
   volume = length × width × height
   $= 3 \text{ feet} \times 0.\overline{6} \text{ feet} \times 1 \text{ foot} = \boxed{2 \text{ ft}^3}$

7. **(B)** $V = \tfrac{4}{3}\pi r^3 = \tfrac{4}{3} \times 3.14 \times (3 \text{ cm})^3 = \boxed{113.04 \text{ cc}}$

8. **(C)** $100 \text{ gallons}\left(\dfrac{2}{5}\right) = \boxed{40 \text{ gallons}}$

9. **(D)** $1,260 \text{ ml.}\left(\dfrac{2}{3}\right) = \boxed{840 \text{ ml.}}$

10. **(A)** $40 \ \cancel{\text{students}} \left(\dfrac{7 \text{ women}}{8 \ \cancel{\text{students}}}\right) = \boxed{35 \text{ women}}$

11. **(A)** $8 \ \cancel{\text{ounces}} \left(\dfrac{1 \text{ quart}}{32 \ \cancel{\text{ounces}}}\right) = \dfrac{8}{32} \text{ quarts} = \boxed{\dfrac{1}{4} \text{ quarts}}$

12. **(C)** 75 students − 15 men = 60 women

$$\frac{60 \text{ women}}{15 \text{ men}} = \boxed{\frac{4}{1}}$$

13. **(D)** $$\frac{5 \text{ ounces}}{5 \text{ ounces} + 15 \text{ ounces}} = \frac{5 \text{ ounces}}{20 \text{ ounces}} = \boxed{\frac{1}{4}}$$

14. **(B)** $8\left(\frac{1}{8}\text{grain}\right) = (1 \text{ tablet})8$

1 grain = 8 tablets

$$1\frac{1}{2}\text{tablets}\left(\frac{1 \text{ grain}}{8 \text{ tablets}}\right) = \frac{1.5}{8}\text{grains}\left(\frac{2}{2}\right) = \boxed{\frac{3}{16}}\text{grains}$$

15. **(B)** Density and volume have an inverse relationship. Gold is more dense than iron; therefore, a 50 g sample of gold will occupy less volume than a 50 g sample of iron.

$$50 \text{ g Au}\left(\frac{1 \text{ cm}^3}{19.3 \text{ g}}\right) = 2.6 \text{ cm}^3 \text{ Au} \qquad 50 \text{ g Fe}\left(\frac{1 \text{ cm}^3}{7.9 \text{ g}}\right) = 6.3 \text{ cm}^3 \text{ Fe}$$

16. **(C)** $$\frac{\$90 - \$80}{\$80} \times 100 = \frac{\$10}{\$80} \times 100 = \boxed{12.5\%}$$

17. **(A)** $$\frac{72}{80} \times 100 = \frac{9}{10} \times 100 = \boxed{90\%}$$

18. **(B)** $\$3,000 \times 20\% = \$3,000 \times 0.20 = \$600$

$\$3,000 - \$600 = \boxed{\$2,400}$

19. **(D)** $$\frac{\$23,650 - \$21,500}{\$21,500} \times 100 = \frac{\$2,150}{\$21,500} \times 100 = \boxed{10\%}$$

20. **(D)** $100\% - 16\frac{2}{3}\% = 83\frac{1}{3}\%$

$\left(83\frac{1}{3}\%\right)(\text{original price}) = \$35$

$$\text{original price} = \frac{\$35}{83\frac{1}{3}\%} = \frac{\$35}{0.8\overline{3}} = \boxed{\$42}$$

21. **(C)** $\$225,000 \times 13.5\% = \$225,000 \times 0.135 = \boxed{\$30,375}$

22. **(C)** $(\text{total sales})(14\%) = \$140$

$$\text{total sales} = \frac{\$140}{14\%} = \frac{\$140}{0.14} = \boxed{\$1,000}$$

23. **(A)** $\$280 \times 20\% = \$280 \times 0.20 = \$56$

$\$280 - \$56 = \boxed{\$224}$

24. **(D)** $$\frac{\$65 - \$50}{\$50} \times 100 = \frac{\$15}{\$50} \times 100 = \boxed{30\%}$$

25. **(D)** $24\% \times 50 \text{ students} = 0.24 \times 50 \text{ students} = 12 \text{ students}$

$50 \text{ students} - 12 \text{ students} = \boxed{38 \text{ students}}$

26. **(C)** $0.2\%(400) = 0.002(400) = \boxed{0.8}$

27. **(C)** $\dfrac{60°}{90° + 30° + 60° + 45° + 75° + 20° + 40°} \times 100 = \dfrac{60°}{360°} \times 100 = \boxed{17\%}$

28. **(B)** $\dfrac{12}{20} \times 100 = \boxed{60\%}$

29. **(C)** $3\% \times 900 \text{ students} = 0.03 \times 900 \text{ students} = 27 \text{ students}$

    $900 \text{ students} - 27 \text{ students} = \boxed{873 \text{ students}}$

30. **(C)** $1\% \times \$23,000 = 0.01 \times \$23,000 = \boxed{\$230}$

31. **(D)** $\dfrac{\$150 - \$120}{\$150} \times 100 = \dfrac{\$30}{\$150} \times 100 = \boxed{20\%}$

32. **(A)** $\text{C } 6 \times 12 = 72 \qquad \text{H } 1 \times 12 = 12 \qquad \text{O } 6 \times 16 = 96$

    $C_6 H_{12} O_6 = 72 + 12 + 96 = 180$

    $\dfrac{96}{180} \times 100 = \boxed{53\%}$

33. **(C)** $3 \text{ ounces} \left(\dfrac{200 \text{ calories}}{8 \text{ ounces}} \right) = 75 \text{ calories}$

    $\dfrac{75 \text{ calories}}{200 \text{ calories}} \times 100 = \boxed{37.5\%}$

34. **(B)** $7 \text{ hours} \left(\dfrac{80 \text{ cards}}{1 \text{ hour}} \right) = 560 \text{ cards filed}$

    $800 \text{ cards} - 560 \text{ cards} = \boxed{240 \text{ cards}}$

35. **(B)** $1 \text{ week} \left(\dfrac{7 \text{ days}}{1 \text{ week}} \right) \left(\dfrac{45 \text{ min exercise}}{1 \text{ day}} \right) \left(\dfrac{1 \text{ hour}}{60 \text{ min}} \right) = \boxed{5.25 \text{ hours}}$

36. **(C)** $80 \text{ miles} \left(\dfrac{1 \text{ hour}}{40 \text{ miles}} \right) = \boxed{2 \text{ hours}}$

37. **(A)** $1 \text{ square yard} \left(\dfrac{\$1,000}{100 \text{ square yards}} \right) = \boxed{\$10}$

38. **(C)** $100 \text{ square feet} \left(\dfrac{\$0.75}{1 \text{ square foot}} \right) = \boxed{\$75}$

39. **(A)** $100 \text{ calories} \left(\dfrac{15 \text{ min}}{60 \text{ calories}} \right) = \boxed{25 \text{ min}}$

40. **(A)** $1 \text{ ounce} \left(\dfrac{1 \text{ kg}}{35 \text{ ounces}} \right) \left(\dfrac{1000 \text{ g}}{1 \text{ kg}} \right) = 28.57 \text{ g} \approx \boxed{29 \text{ g}}$

41. **(B)**

$$90 \text{ mg} \left(\frac{1 \text{ hour}}{24 \text{ mg}} \right) = 3.75 \text{ hours}$$

$$0.75 \text{ hour} \left(\frac{60 \text{ min}}{1 \text{ hour}} \right) = 45 \text{ min}$$

$$3.75 \text{ hours} = \boxed{3 \text{ hours } 45 \text{ min}}$$

42. **(A)**

$$0.5 \text{ min} \left(\frac{60 \text{ sec}}{1 \text{ min}} \right) \left(\frac{1,100 \text{ feet}}{1 \text{ sec}} \right) \left(\frac{1 \text{ mile}}{5280 \text{ feet}} \right) = 6.25 \text{ miles} \approx \boxed{6 \text{ miles}}$$

43. **(B)**

$$8 \left(\tfrac{5}{8} \text{ mile} \right) = (1 \text{ kilometer}) 8$$

$$5 \text{ miles} = 8 \text{ kilometers}$$

$$2 \text{ miles} \left(\frac{8 \text{ kilometers}}{5 \text{ miles}} \right) = \boxed{3.2 \text{ kilometers}}$$

44. **(C)**

$$5,500 \text{ bandages} \left(\frac{\$50}{1,000 \text{ bandages}} \right) = \boxed{\$275}$$

45. **(A)**

$$228 \text{ watches} \left(\frac{\$1,250}{100 \text{ watches}} \right) = \boxed{\$2,850}$$

46. **(B)**

$$\$76.50 \left(\frac{100 \text{ cents}}{\$1} \right) \left(\frac{1 \text{ yard}}{42\frac{1}{2} \text{ cents}} \right) = \boxed{180 \text{ yards}}$$

47. **(B)**

$$392 \text{ calories} \left(\frac{4 \text{ ounces}}{448 \text{ calories}} \right) = \boxed{3.5 \text{ ounces}}$$

48. **(D)**

$$1 \text{ dozen} \left(\frac{12 \text{ hankies}}{1 \text{ dozen}} \right) \left(\frac{\$1.29}{3 \text{ hankies}} \right) = \boxed{\$5.16}$$

49. **(C)**

$$345 \text{ pins} \left(\frac{\$4.15}{100 \text{ pins}} \right) = \boxed{\$14.32}$$

50. **(D)**

$$2.5 \text{ gallons punch} \left(\frac{1 \text{ pint grape juice}}{0.5 \text{ gallons punch}} \right) \left(\frac{16 \text{ ounces grape juice}}{1 \text{ pint grape juice}} \right) \left(\frac{1 \text{ quart grape juice}}{32 \text{ ounces grape juice}} \right) = \boxed{2.5 \text{ quarts}}$$

51. **(C)**

$$4 \left(\tfrac{3}{4} \text{ gram} \right) = (1 \text{ tablet}) 4$$

$$3 \text{ grams} = 4 \text{ tablets}$$

$$4.5 \text{ g} \left(\frac{4 \text{ tablets}}{3 \text{ g}} \right) = \boxed{6 \text{ tablets}}$$

52. **(C)**

$$36 \text{ calories} \left(\frac{1 \text{ gram}}{4 \text{ calories}} \right) = \boxed{9 \text{ grams}}$$

53. **(B)**

$$1 \text{ borough} \left(\frac{245 \text{ sections}}{5 \text{ boroughs}} \right) = \boxed{49 \text{ sections}}$$

54. **(D)**

$$1 \text{ plow} \left(\frac{45 \text{ miles}}{9 \text{ plows}} \right) = \boxed{5 \text{ miles}}$$

55. **(A)**

$$10 \text{ days} \left(\frac{5 \text{ min}}{1 \text{ day}} \right) = \boxed{50 \text{ min}}$$

56. **(C)**

$$28 \text{ miles} \left(\frac{1 \text{ hour}}{3 \text{ miles}} \right) = 9\tfrac{1}{3} \text{ hour}$$

$$\tfrac{1}{3} \text{ hour} \left(\frac{60 \text{ min}}{1 \text{ hour}} \right) = 20 \text{ min}$$

$$9\tfrac{1}{3} \text{ hour} = \boxed{9 \text{ hour } 20 \text{ min}}$$

57. **(A)**

$$42 \text{ small revolutions} \left(\frac{8 \text{ cogs}}{1 \text{ small revolution}} \right) \left(\frac{1 \text{ large revolution}}{24 \text{ cogs}} \right) = \boxed{14 \text{ large revolutions}}$$

58. **(D)**

$$10\tfrac{1}{2} \text{ pounds} \left(\frac{1 \text{ basket}}{1\tfrac{1}{2} \text{ pounds}} \right) = \boxed{7 \text{ baskets}}$$

59. **(D)**

$$\$61.60 \left(\frac{1 \text{ tee shirt}}{\$5.60} \right) = \boxed{11 \text{ tee shirts}}$$

60. **(B)** The freezing point of water is 32°F and the boiling point of water is 212°F. The difference between the boiling point and the freezing point of water is 180°F.

$$212°F - 32°F = \boxed{180°F}$$

The number of degrees between the freezing point and the boiling point of water on the Celsius or Kelvin scales is 100 degrees.

In-Depth Solutions for Matter and The Periodic Table

1. **The correct answer is (B).** Classifies and categorizes both mean "separates into groups with like properties."

2. **The correct answer is (A).** A theory is most similar to a principle because both require supporting evidence.

3. **The correct answer is (D).** Heterogeneous is the antonym of homogeneous. Homogeneous means "having similar structure because of common decent," while heterogeneous means "differing or opposing in structure."

4. **The correct answer is (A).** Negative is the antonym of positive. The charge on an electron is negative and the charge on a proton is positive. These subatomic particles have opposite charges.

5. **The correct answer is (C).** Stubborn is the antonym of ductile. Ductile means easily stretched, while stubborn means refusing to yield or hard to handle.

6. **The correct answer is (A).** Emit is the antonym of absorb. Absorb means to take in, while emit means to give off.

7. **The correct answer is (A).** The nurse had to weigh the baby. Weigh means to find the weight of, while way means a path or a means of passing from one place to another.

8. **The correct answer is (B).** She wanted to purchase two quarts of milk. Quart is a unit of measure, while quartz is a mineral.

9. **The correct answer is (B).** A calorie is the unit of measure of heat—the amount of heat required to raise the temperature of 1 gram of water by 1°C.

10. **The correct answer is (B).** Physical properties are those that do not involve any change in the nature or chemical composition of the substance. Heating a substance to the temperature at which it boils (its *boiling point*) will change its physical state but will not alter its chemical state.

11. **The correct answer is (A).** Ozone is a molecule variety of oxygen in which 3 oxygen atoms bond to make 1 molecule of O_3. Ozone is easily formed by the action of electricity or ultraviolet radiation on the normal diatomic form (O_2) of oxygen.

12. **The correct answer is (B).** Water is a compound formed by oxygen and hydrogen in chemical combination. Blood and air are mixtures, and oxygen is an element.

13. **The correct answer is (D).** Emergent properties of matter cause atoms that react to form compounds to lose their original properties and gain properties of the new substance.

14. **The correct answer is (B).** Argon, barium, and aluminum are in groups VIIIA, IIA, and IIIA, respectively. Copper is the only element that is in the transition area of the periodic table (group IB).

15. **The correct answer is (C).** Transition elements are characterized by atoms in which the two highest energy levels are incompletely filled. Consulting the periodic table indicates that of those listed, only nickel is a transition element (see the periodic table).

16. **The correct answer is (A).** The atoms of each element in group Zero have 8 electrons (except helium) in the outermost energy level. The elements have atoms with complete octets of electrons, producing stable energy levels.

17. **The correct answer is (A).** Sodium (Na) and potassium (K) are both in group IA, with 1 electron in the outer energy level. Sodium and potassium have three and four energy levels, respectively.

18. **The correct answer is (B).** Reactivity of elements in group IA increases from top to bottom because valence electrons are progressively further from the positive charges in the nucleus of atoms. Reactivity of elements in group VIIA increases from bottom to top because valence electrons are progressively closer to the positive charges of the nucleus.

19. **The correct answer is (A).** Sodium (Na) is a metal in group IA, with 1.0 electron in its outermost shell. Its most stable configuration is achieved by donating 1 electron, resulting in a + 1 valence.

20. **The correct answer is (B).** The noble gases have stable electronic structures with eight outer "valence" electrons (2 s and 6 p)—excluding helium (which has a complete valence shell with only 2 s electrons)—and therefore have no compulsion to seek means of donating, accepting, or sharing electrons. The halogens need an electron to achieve stability, whereas group IIA need to donate two electrons to form stable ions with eight electrons on the outermost shell. Group IB metals react but show a high resistance to oxidation.

21. **The correct answer is (A).** Hydrogen is the lightest element known, having an atomic weight of 1.008. The atom contains one proton, no neutrons, and one electron.

22. **The correct answer is (B).** The chemical activity of an atom is how it reacts with other atoms. Atoms are governed by the number of electrons in their outermost shell. Helium, neon, and other atoms with no electron vacancies in their outermost shell are inert; they tend not to enter into chemical reactions. Hydrogen, oxygen, and other atoms with electron vacancies in their outermost shell tend to interact with other atoms.

23. **The correct answer is (A).** Within a group, the element with the lowest atomic number has the highest ionization energy.

24. **The correct answer is (B).** The atomic radii of a group of metals increase as the atomic number increases. Therefore, the element with the lowest atomic number has the smallest atomic radius.

25. **The correct answer is (D).** Within a group of nonmetals such as the halogens, the one with the highest atomic number has the least electronegative value.

26. **The correct answer is (A).** Electronegativism increases within a group in the periodic table as the atomic number decreases. The example shown here is the progression from iodine to fluorine.

27. **The correct answer is (C).** Phosphorus has fifteen electrons, all of which are paired except for three. The three electrons in the 3p subshell are unpaired.

28. **The correct answer is (C).** Chlorine is in the third period and has seven electrons in its outermost shell. Two electrons are in an s orbital, leaving five electrons for the p orbitals.

29. **The correct answer is (B).** Sulfur in group VIA needs two electrons added to its six to achieve a stable outer shell of eight electrons.

30. **The correct answer is (B).** For many atoms, the simplest way to attain a completely filled outer energy level is either to gain or to lose one or two electrons.

31. **The correct answer is (C).** The 4 oxygen ions would have a total *valence* of –8 (each oxygen, –2); if the valence of the SO_4 ion is –2, the remaining valence of 6 would be matched to a positive valence. Therefore, sulfur must have a valence of + 6.

32. **The correct answer is (D).** Sulfuric acid (H_2SO_4) contains 4 oxygens, each with a valence of –2; the total negative valence is –8. Each of the 2 hydrogens has a valence of + 1; the total hydrogen valence equals + 2. Therefore, sulfur would need a valence of + 6 to equalize the positive and negative valences in the molecule.

33. **The correct answer is (A).** Sulfuric acid is composed of 2 hydrogen ions and 1 sulfate ion. The chemical formula is H_2SO_4; thus, its chemical name is hydrogen sulfate.

34. **The correct answer is (B).** Sodium bisulfate differs from sodium sulfate in having hydrogen as a part of the molecule. Since the sulfate ion has a valence of –2, and hydrogen and sodium each have a valence of + 1, then one sodium and one hydrogen must be in combination with the sulfate ion ($NaHSO_4$).

35. **The correct answer is (C).** The NH_4^+ species has a positive valence, which has affinity for a negative chloride (Cl-) ion.

36. **The correct answer is (C).** A hydrogen atom must accept an electron in order to form a hydride ion. No other changes occur. The hydride therefore has a single negative charge, whereas the hydrogen atom is neutral. An extra negative charge with no change in positive charge expands the radius. Consequently, the hydrogen atom and the hydride atom both have the same number of protons.

37. **The correct answer is (D).** In the series for the oxyhalogen ions (e.g., ClO_4^-, ClO_3^-, ClO_2^-, ClO^-), the one with the least amount of oxygen is designated hypochlorite.

38. **The correct answer is (D).** Hydrogen in molecular form is composed of 2 atoms; the formula for 1 molecule of hydrogen is H_2. Therefore, 2 molecules of hydrogen is $2H_2$.

39. **The correct answer is (B).** Calcium hypochlorite ($Ca(OCl)_2$) is effective as a bleach due to the oxidizing activity of hypochlorous acid (HClO), which is produced from calcium hypochloride, and from which chlorine gas can also be liberated.

40. **The correct answer is (B).** The *atomic number* represents the number of protons within the nucleus of an atom; since an atom is electrically neutral, the atomic number also equals the number of electrons of the atom. The atomic weight (*mass number*) represents the total number of nuclear particles (protons plus neutrons). All atoms of the same element have the same number of protons and, thus, the same atomic number. If the numbers of neutrons differ, the mass number is different; these atoms, then, are *isotopes*.

41. **The correct answer is (B).** The atomic number indicates the number of *protons* (positively charged particles) within the nucleus of an atom. Since an atom is electrically neutral, the number of protons is equal to the number of *electrons* (negatively charged particles) of the atom. Thus, the atomic number can indicate the number of electrons as well as the number of protons.

42. **The correct answer is (D).** The atomic weight (or mass number of an atom) equals the total of the number of protons and neutrons in the nucleus of the atom, or the total number of particles in the nucleus. Electrons have negligible (very small) mass.

43. **The correct answer is (D).** The nucleus of an atom consists of protons and neutrons. A proton is a subatomic particle with positive electric charge. The number of protons in the nucleus of an atom is equal to the atomic number.

44. **The correct answer is (C).** Regardless of the element, atoms have just as many electrons as protons. This means that they carry no net charge, overall. A proton is always positively charged, and an electron is negatively charged.

45. **The correct answer is (D).** The statement given is the definition of isotope, a form of atom of an element that differs from other atoms of the same element by the number of neutrons.

46. **The correct answer is (B).** If the atomic mass of an isotope is unknown, the number of neutrons cannot be determined. However, if the atomic number is known, the number of protons is known. In an atom, there is no charge, so the number of electrons must equal the number of protons and the atomic number.

47. **The correct answer is (A).** The mass number of an isotope is the sum of the protons and neutrons in an atom. Seventeen protons and seventeen neutrons are necessary to give a mass number of thirty-four. The number of electrons equals the number of protons in an atom.

48. **The correct answer is (C).** Gram molecular weight may be defined as the weight of the molecule expressed in grams. The molecular weight of glucose ($C_6H_{12}O_6$) is 180, as the following indicates: The weight of the six atoms of carbon equals 72 (6×12); twelve hydrogens weigh 12 (12×1); the weight of the six oxygens equals 96 (6×16). The total is 180; thus, the gram molecular weight of glucose is 180 grams.

49. **The correct answer is (A).** The molecular weight of sodium hydroxide (NaOH) is 40×1 or 40. This can be calculated by adding the atomic weights of the atoms composing the molecule.

50. **The correct answer is (C).** Of the 17 atoms making up a molecule of $Al_2(SO_4)_3$, there are two atoms of aluminum, three atoms of sulfur, and twelve atoms of oxygen (four oxygens in each of the three sulfate ions). The weight of the oxygen is 192; the total weight of the molecule is 342. Thus, the percentage of oxygen by weight is $\frac{192}{342}$, or 56 percent.

51. **The correct answer is (D).** For a molecular compound, the gram formula weight is the same as the gram molecular weight. The empirical formula represents the smallest ratio of the different atoms possible in the compound. The molecular formula may be a multiple of the empirical formulas (e.g., twice).

In-Depth Solutions for Bonding, Structure, and Reactions

1. **The correct answer is (C).** An ionic bond will form between 2 atoms if the loss of electrons from one to the other will result in both atoms having a completely filled outer electron orbit. A sodium atom has 11 electrons, 2 in its first electron orbit, 8 in its second electron orbit, and 1 in its outer electron orbit. The loss of 1 electron from the sodium atom will eliminate its outer electron orbit, making the next orbit in toward the nucleus the new outer orbit with a full complement of 8 electrons. A chlorine atom has 17 electrons, 2 in its first electron orbit, 8 in its second, and 7 in its outer electron orbit. The acceptance of 1 electron by the chlorine atom will fill its outer electron orbit with the full complement of 8 electrons.

2. **The correct answer is (C).** The electrons in the outer energy level comprise the portion of an atom directly involved in ionic bonding. The outer energy level will either gain or lose electrons in order to achieve a complete octet (8 electrons).

3. **The correct answer is (B).** Covalence is the mutual attraction between two atoms that share a pair of electrons. It is the strongest type of chemical bond.

4. **The correct answer is (A).** In polar covalent bonds, atoms of different elements, which have a different number of protons, do not exert the same pull on shared electrons. The more attractive atom ends up with a slight negative charge; the atom is "electronegative." Its effect is balanced out by the other atoms, resulting in a slight positive charge. In simple words, a polar covalent bond has no net charge-but the charge is distributed unevenly between the bond's two ends.

5. **The correct answer is (B).** In a hydrogen bond, a small, highly electronegative atom of a molecule interacts weakly with a hydrogen atom that is already participating in a polar covalent bond.

6. **The correct answer is (C).** Calcium and the other metals of the group IIA ionize by losing two electrons. It achieves the stable electron structure of argon, with eight electrons in the outer shell.

7. **The correct answer is (C).** Molecular solids are held together by rather weak London forces, which increase with molecular weight. The melting points of molecular substances are relatively low since there are no strong forces, such as ionic forces, to overcome.

8. **The correct answer is (D).** Ionic compounds form readily between non-metals and metals. The only set that has no such combination is CH_2O, H_2S, and NH_3.

9. **The correct answer is (D).** The electrical charge results when a neutral atom or group of atoms loses or gains one or more electrons during chemical reactions.

10. **The correct answer is (C).** Distortion is the antonym of symmetry. Symmetry means similarity of form on either side of a dividing line. Distortion means twisted out of shape.

11. **The correct answer is (C).** An atom lying at the corner of a unit cell will touch four corners of cubes below and four corners of cubes above.

12. **The correct answer is (D).** Bronze is an alloy of copper and tin. An alloy is made by mixing two or more metals while in molten condition and then allowing the mixture to cool and solidify. The metals remain completely dissolved in one another after solidification, forming a homogenous substance with properties different from any of the constituent metals in their pure forms. For example, bronze is both harder and more resistant to corrosion than pure copper.

13. **The correct answer is (A).** Heating water to a very high temperature (about 2,000°C) will cause only about 2 percent of the water to dissociate into hydrogen and oxygen—not a very economical or efficient way of obtaining oxygen. Electrolysis will decompose water at a much lower temperature. Hydrogen peroxide and many oxides or metals can be decomposed more easily by heating alone.

14. **The correct answer is (D).** Only one substance (HCl) is dissociated; therefore, the reaction is a single displacement.

15. **The correct answer is (A).** Zinc is ranked above the others but below aluminum in the activity series of metals. The higher the metal is ranked, the more energetic it is in its displacement ability. Therefore, aluminum could displace zinc in the zinc chloride solution, forming aluminum chloride. For this reason, zinc chloride should not be stored in a tank made of aluminum because aluminum will displace the zinc in the zinc chloride.

16. **The correct answer is (B).** Carbon disulfide is highly inflammable; its complete combustion yields carbon dioxide and sulfur dioxide.

17. **The correct answer is (D).** Because fluorine is extremely active, it does not occur freely in nature but combines naturally with all elements except the inert gases, forming very stable compounds. Because it is a vigorous oxidizing agent, it cannot be oxidized by other oxidizing agents. For these reasons, fluorides are more difficult to decompose than are compounds of the other halogens.

18. **The correct answer is (C).** Charcoal is an amorphous form of carbon, and carbon can react with oxides of many metals and other substances to form CO, CO_2, and the carbides of the metals. Thus, the oxides would be reduced by the removal of oxygen, and carbon would be oxidized to carbon monoxide or carbon dioxide; therefore, oxidation and reduction would occur.

19. **The correct answer is (B).** Oxidation-reduction reactions are those in which one substance is oxidized by the loss of electrons, and another substance is reduced by the gain of electrons. In double replacement reactions, this does not occur. Ions are exchanged between reactants and do not lose or gain electrons.

20. **The correct answer is (A).** For every oxidation, there must be a reduction. Oxidation can involve the removal of electrons, the removal of hydrogen, or the addition of oxygen; reduction, the opposite. In equation (A) metallic sodium loses an electron to become a positively charged ion; chlorine gains an electron to become a negatively charged chloride ion. Thus, sodium is oxidized and chlorine is reduced.

21. **The correct answer is (A).** Since the oxidation number of a compound is 0, the total positive values must equal the total negative values. Consequently, four times the value for oxygen (-2) must equal the positive values due to potassium (+1 each) and manganese (Mn).

$$4(-2) + 1(+1) + ? = 0$$
$$-8 + 1 + ? = 0$$
$$? = +7$$

22. **The correct answer is (B).** Although hydrogen is relatively inert at ordinary temperatures, it can combine with free oxygen, with oxygen that is in chemical combination, with some metallic elements, and with many nonmetallic elements. The addition of hydrogen reduces a substance.

23. **The correct answer is (A).** All acids contain hydrogen, which may be displaced by certain metals. However, copper ranks below hydrogen in the *electromotive* or *activity series*. *Activity* refers to the activity of a metal in displacing hydrogen in acids and in water. Metals ranked below hydrogen in the series do not displace hydrogen from acids.

24. **The correct answer is (D).** Hydrochloric acid reacts with active metals, forming the chloride of that metal and releasing hydrogen.

25. **The correct answer is (D).** Free carbon dioxide and water will combine to produce carbonic acid (H_2CO_3). Different reactants would be needed to produce ozone (O_3), methane (CH_4), or hydrogen peroxide (H_2O_2).

26. **The correct answer is (B).** Hydrochloric acid can react with sodium hydrogen sulfite to produce sodium chloride, water, and sulfur dioxide. Sulfur dioxide can react with barium hydroxide to form barium sulfite and water.

27. **The correct answer is (C).** Nitric acid reacts with the base potassium hydroxide to form potassium nitrate and water. For each hydrogen ion of the acid, there is a hydroxyl ion of the base to combine with it to form water. Therefore, neutralization occurs.

28. **The correct answer is (D).** Potassium hydroxide (KOH) and hydrochloric acid (HCl) react to form potassium chloride (KCl) and water. This is a double displacement.

29. **The correct answer is (D).** The H^+ has no electrons, which enables it to use two unshared electrons from N to form a coordinate covalent bond.

30. **The correct answer is (B).** The salt that would yield a weak acid when mixed with acid would dissolve most readily. Calcium carbonate yields carbonic acid (H_2CO_3) when it is acidified. The other salts yield hydrochloric acid (HCl), nitric acid (HNO_3), or phosphoric acid (H_3PO_4), which are strong acids.

31. **The correct answer is (B).** The catalytic converter uses platinum to convert carbon monoxide and hydrocarbons to carbon dioxide and water, resulting in reduced air pollution.

32. **The correct answer is (D).** Equation (D) is balanced, because the amount of each substance is equal on both sides of the equation. There are two mercury atoms and two oxygen atoms on each side.

33.   **The correct answer is (B).** In a chemical reaction, the quantities of reactants/products are directly proportional.

Let $x_1 = 2$ moles of A and $y_1 = 5$ moles of B.

Then $x_2 =$ number of moles of A and $y_2 = 7$ moles of B such that:

$$\frac{5}{2} = \frac{7}{x_2}; \ x_2 = 2.8 \text{ moles of A.}$$

34. **The correct answer is (A).** Magnesium has an atomic weight of 24; hydrogen's atomic weight is approximately 1. As is true with many metals, magnesium can react with an acid, such as hydrochloric acid, to produce hydrogen ($Mg + 2HCl \rightarrow MgCl_2 + H_2$). The amount of hydrogen produced can be calculated by using the following equation:

$$6 \text{ g Mg} \left(\frac{1 \text{ mole Mg}}{24 \text{ g Mg}} \right) \left(\frac{1 \text{ mole } H_2}{1 \text{ mole Mg}} \right) \left(\frac{2 \text{ g } H_2}{1 \text{ mole } H_2} \right) = 0.5 \text{ g } H_2$$

35. **The correct answer is (D).** Since there is twice as much hydrogen as there is oxygen in a molecule of water (H_2O), the 26 milliliters of hydrogen could combine with only 13 milliliters of oxygen. This would leave 11 of the 24 milliliters of oxygen uncombined.

36. **The correct answer is (C).** The number of moles of HCl required for the titration must equal the number of moles of NaOH. Therefore 20 ml of 0.4 ml Cl is required. After titration, the total volume of the mixture is 40 + 20 = 60 ml.

$$40.0 \text{ mL NaOH} \left(\frac{1 \text{ L NaOH}}{1000 \text{ mL NaOH}} \right) \left(\frac{0.20 \text{ moles NaOH}}{1 \text{ L NaOH}} \right) \left(\frac{1 \text{ mole HCl}}{1 \text{ mole NaOH}} \right) \left(\frac{1 \text{ L HCl}}{0.4 \text{ moles HCl}} \right) \left(\frac{1000 \text{ mL HCl}}{1 \text{ L HCl}} \right) = 20 \text{ mL HCl}$$

37. **The correct answer is (C).** The amount of heat absorbed or lost in a chemical reaction is the heat of reaction. It equals the heat content of the product(s) minus the heat content of the reactants. Heat is absorbed if more energy is stored in the products than in the reactants (a positive heat of reaction), and released if more energy is stored in the reactants than in the product(s) (a negative heat of reaction). In the formation of a compound, the heat of reaction is called the heat of formation for that compound. A compound with a high heat of formation has more energy stored in it than in the reactants and is difficult to form and easy to decompose.

38. **The correct answer is (C).** The reactants of some reactions have more energy than the products. Many such reactions release energy that cells can use, which is what happens during aerobic respiration.

39. **The correct answer is (C).** Exergonic reactions are downhill reactions with negative energy changes because the products have less energy than the reactants.

40. **The correct answer is (C).** The human body is an open system—exchanging nutrients, CO_2 and O_2, nitrogenous waste, water, and heat with the external environment.

41. **The correct answer is (B).** Chemical kinetics is the study of reaction rates and how these change with variation of conditions along with the various molecular events that transpire during the overall reaction.

42. **The correct answer is (A).** A first order reaction has a rate that is proportional to the concentration of only one reactant.

43. **The correct answer is (A).** At the initiation of the reaction, all substances are reactants (A and B).

44. **The correct answer is (A).** The affinity of A and B is greater than that of C and D, resulting in a greater concentration of product than reactants at equilibrium.

45. **The correct answer is (D).** The reaction is shifted to the left.

46. **The correct answer is (B).** The arrows directed to the right in chemical equations represent products generated from the forward reaction, and those directed to the left represent reactants produced by the reverse reaction. The arrow length can be used to symbolize the concentration of the substances on both sides of the equation.

47. **The correct answer is (C).** The total number of moles of reactants and products are equal; therefore, a change of pressure has no effect in the equilibrium. Increasing hydrogen or reducing hydrogen bromide drives the reaction to the right. Only increasing hydrogen bromide shifts the equilibrium to the left.

48. **The correct answer is (A).** Heat is produced by the reaction, and, according to Le Chatalier's Principle, the system will adjust to maintain equilibrium. Heating would force the reaction to the left. A pressure increase shifts the reaction in the direction of smaller volumes (right), and addition of more reactant (N_2 or H_2) also pushes the reaction to the right.

In-Depth Solutions for Gases, Boiling, and Specific Heat

1. **The correct answer is (A).** Air is a mixture of gases; the gases present in greatest quantity are nitrogen and oxygen. The quantity of nitrogen in the air is almost four times that of oxygen. Of the compounds listed, methane (CH_4), with a molecular weight of 16, is lighter than air.

2. **The correct answer is (B).** Carbon dioxide is an odorless gas that is heavier than air. It is colorless, soluble in water, and about one and one-half times as heavy as air. It is a component of air, but comprises less than 0.05 percent of it.

3. **The correct answer is (C).** The barometer is an instrument used to measure air pressure and to forecast weather.

4. **The correct answer is (B).** Air pressure at sea level is 14.7 pounds/square inch. This pressure, at $0°$ C supports a barometer mercury column of 76 centimeters, or 30 inches. This is equivalent to one atmosphere of pressure.

5. **The correct answer is (B).** The atmosphere is composed of about 78 percent nitrogen.

6. **The correct answer is (B).** Nitrogen composes nearly four fifths of the air; oxygen nearly one fifth. The remainder of air consists of other gases.

7. **The correct answer is (D).** Air is a mixture of gases; dry air consists of about 78 percent nitrogen and 21 percent oxygen. The remaining 1 percent consists of other gases, including carbon dioxide.

8. **The correct answer is (A).** Air is a mixture of gases; dry air consists of about 20 percent oxygen.

9. **The correct answer is (D).** The pressure of a gas is measured by the distance in millimeters that it will lift a column of mercury in a barometer. The pressure of air is 760 mm of mercury at sea level. Air is a mixture of gases, and the pressure of each gas in such a mixture, referred to as its *partial pressure*, is equal to its concentration in the mixture. Oxygen constitutes 21% of air; therefore, its partial pressure at sea level would be 21% of 760 mm of mercury of pressure, or 160 mm of mercury.

10. **The correct answer is (D).** Nitrogen makes up about 80 percent of the air, and therefore, contributes about 80 percent of the total atmospheric pressure.

$$0.80 \times 760mm = 608mm$$

11. **The correct answer is (B).** *Evaporation* is the physical change of a substance from a liquid to a gas. The rate of movement of gas molecules into the air is more rapid in dry air or when the humidity is low, and decreases with increased humidity. The movement of air currents (wind) carries away the vapor above the surface of the liquid, thus increasing evaporation rate.

12. **The correct answer is (A).** Gases are affected by temperature and pressure. If the gas is in a closed system, it cannot change its volume if temperature or pressure changes. Therefore, in a closed system, increasing the pressure would increase the temperature.

13. **The correct answer is (B).** Boyle's Law illustrates the relationship between the pressure and volume of a mass of gas at a fixed temperature. The law can be simply stated: at a given fixed temperature and mass of gas, pressure, and volume are inversely proportional. For example, volume will increase as pressure decreases, or volume will decrease as pressure increases. Therefore, pressure × volume = a constant.

14. **The correct answer is (A).** The inhaling and the exhaling of air by the human lungs demonstrates Boyle's Law. Inhaling results from the expansion of the chest cavity (increase in volume and decrease in air pressure) and exhaling results from compression of the chest cavity (decrease in volume and increase in air pressure). These changes are accomplished by movement of the ribs and diaphragm.

15. **The correct answer is (B).** Two parameters are directly proportional if their ratio is a mathematical constant ($K = \dfrac{x}{y}$).

16. **The correct answer is (B).** Since all three gases are in equal size flasks with the same temperature and pressure, the total number of moles (hence, the total number of molecules) is equal in each case. However, a molecule of NH_3 contains four atoms, a molecule of NO contains two atoms, and a molecule of N_2 contains two atoms.

17. **The correct answer is (D).** One mole (gram molecular weight) of nitrogen is 28 grams, since two nitrogen atoms form a nitrogen molecule (N_2). One mole of gas occupies a volume of 22.4 liters at standard conditions.

18. **The correct answer is (A).** One mole (*gram molecular weight*) of a gas occupies 22.4 liters under standard conditions. Since 1 liter of the gas in question, we can set up the equation as follows:

$$1\,L\left(\frac{1\,\text{mole}}{22.4\,L}\right) = 0.045\ \text{moles}$$

$$\text{Molar mass} = \frac{\text{mass}}{\text{moles}} = \frac{1.16\,g}{0.045\,\text{moles}} = 26\,\frac{g}{\text{mole}}$$

$$C_2H_2 = 2\times12+2\times1 = 26;\ CO = 1\times12+1\times16 = 28;$$
$$NH_3 = 1\times14+3\times1 = 17;\ O_2 = 2\times16 = 32$$

C_2H_2 is the gas with the molar mass that matches the calculated value.

19. **The correct answer is (D).** The ratio of the rates of effusion of two gases (proportional to the speeds of the gas molecules) is inversely proportional to the square roots of their molecular weights; therefore, if gas A 4 times heavier than gas B, gas A has an average speed that is half of the average speed of gas B. Or rephrased, gas B has twice the average speed of gas A.

$$\frac{\text{rate A}}{\text{rate B}} = \sqrt{\frac{\text{mass B}}{\text{mass A}}} = \sqrt{\frac{\text{mass B}}{4(\text{mass B})}} = \sqrt{\frac{1}{4}} = \frac{1}{2}$$

$$\frac{\text{rate A}}{\text{rate B}} = \frac{1}{2} \quad \Rightarrow \quad \text{rate B} = 2\big(\text{rate A}\big)$$

20. **The correct answer is (D).** *Potential energy* is, in effect, energy in storage that can be released when conditions are conducive. *Kinetic energy* is energy of motion.

21. **The correct answer is (C).** The stored form of energy is called potential energy. An example is the glucose molecule; when it breaks down, it releases a large quantity of chemical energy to do work.

22. **The correct answer is (A).** A higher boiling point for a liquid is seen when more energy is needed to separate the gaseous molecules from the liquid. It shows that for the same amount of energy, a higher boiling liquid would expel fewer molecules as vapor at any temperature, resulting in a lower vapor pressure. The lowest boiling substance vaporizes most easily, and therefore, would have the highest vapor pressure at a specific temperature.

23. **The correct answer is (D).** Fractional distillation can be used to separate the components of a mixture; the mixture is heated to boiling and the temperature is constantly raised. The component with the lowest boiling point vaporizes first; the component with the highest boiling point will vaporize last. Liquid oxygen has a boiling point of −183°C, which is higher than the boiling point of the other components; therefore, oxygen will boil off last.

24. **The correct answer is (D).** Sulfuric acid has a high boiling point: 317°C. Because of this, it is not very volatile and can be used to prepare more volatile acids, such as hydrochloric and nitric acids.

25. **The correct answer is (A).** The problem can be solved by using the formula: c = Q/MDT. Inasmuch as Q, M, D, and T are given, no algebraic rearrangement is necessary to solve for c (specific heat).

$$c = \frac{3,480 \text{ calories}}{(300 \text{ g})(70°C - 50°C)} = 0.58 \tfrac{\text{cal}}{\text{g} \cdot °C}$$

The value calculated matches most closely with ethyl alcohol.

26. **The correct answer is (A).** One Calorie (kilocalorie) will raise 1 kilogram (1,000 grams) of water 1°C. It would take 18 Calories to raise 1,000 grams of water 18 degrees (38° − 20°), or 36 Calories to raise 2,000 grams of water 18 degrees Celsius.

In-Depth Solutions for Solutions

1. **The correct answer is (C).** A saturated mixture is one in which the solution contains as much solute as will dissolve in the solvent at a given temperature. Saturated in this context does not refer to a mixture; however, the meaning of saturated is "full of" moisture, much like a saturated mixture is "full of" solute.

2. **The correct answer is (B).** The liquid part of the solution is called solvent.

3. **The correct answer is (C).** A *saturated* solution is one that is in equilibrium; a condition of equilibrium exists between the solvent and the solute, and no more of the solute can dissolve.

4. **The correct answer is (D).** A supersaturated solution contains more dissolved solute than is normal at a given temperature. A disturbance will cause the solution to adjust to the normal concentration with the concurrent expulsion of the excess solute. A saturated solution will remain stable.

5. **The correct answer is (B).** As water is warmed, its solubility of oxygen decreases, resulting in a lower O_2 content and producing unfavorable conditions for animals.

6. **The correct answer is (B).** Water is a polar molecule whereas oil is a nonpolar molecule. Water molecules have a slightly negative charge in one side and slightly positive charge in the other side, but lipid molecules do not have any charges.

7. **The correct answer is (D).** The concentration of a solution, on a percentage basis, contains a specific amount of solute in grams per 100 milliliters of solution. Thus, a 10-percent solution would have 10 grams of solute per 100 milliliters of solution.

8. **The correct answer is (C).** A molar solution contains one gram molecular weight (the molecular weight of the molecule expressed in grams), or one mole per liter of solution. Therefore, each is a one-molar solution.

9. **The correct answer is (C).** There is one phosphorus in the formula, but there are three potassiums and four oxygens. The number of atoms or ions present per formula unit equals the corresponding number of moles of each type in one molecule of the compound (the amount present in a liter of a 1 molar solution). In solution, potassium phosphate would separate as potassium ions and phosphate ions.

10. **The correct answer is (C).** A molar solution contains one mole of solute per liter of solution. Therefore, one mole of solute is combined with less than one kilogram (one liter) of water. In a one molal solution, which contains one mole of solute per kilogram of water, the total volume for a solution of one mole exceeds one liter. Therefore, one liter of a one molal solution contains less than one mole.

11. **The correct answer is (B).** A 1.0 molar solution of a substance is prepared by dissolving 1 gram-molecular weight of solution in enough solvent to equal 1 liter of solution. A 1 normal solution is prepared by dissolving an amount of solute (gram-molecular weight/total positive valence) in enough solvent to equal 1 liter of solution. (Atomic weights H = 1.0; S = 32; O = 16). mol. wt. = 98, $\frac{96g}{L} = 1M$ and $\frac{98}{2} = \frac{49g}{L} = 1N$.

12. **The correct answer is (B).** A half of the dosage is 5×10^{-5} M.

13. **The correct answer is (D).** Sugar alone is the only molecular compound of this group. The others readily dissociate as ions in polar solvents, such as water.

14. **The correct answer is (B).** An *electrolyte* is a substance that forms an electrically conducting solution when dissolved in water. This is caused by the dissociation of the compound into ions, or *ionization*. The greater the ionization, the stronger the electrolyte. Calcium chloride ionizes to a much greater degree than the others and is a strong electrolyte.

15. **The correct answer is (B).** The greater the concentration of soluble particles, the greater the lowering of the freezing point of the solvent. A mole of calcium chloride contains one mole of calcium ions and two moles of chloride ions. A mole of sodium chloride contains only two moles of ions, and sucrose and methanol have only one mole of particles each per liter of one molar solution.

16. **The correct answer is (C).** Colligative properties are based on the number of particles. Vapor pressure, osmotic pressure, boiling point elevation, and freezing point lowering all are colligative properties. Density is not.

17. **The correct answer is (C).** Sodium chloride is an ionic substance that dissociates in aqueous solutions, to give 2 times the number of particles per mole as molecular substances such as glucose.

$$\frac{0.315 \text{ M}}{2} = 0.157 \text{ M}$$

18. **The correct answer is (A).** Cell membranes are selective permeable, which determines which substances may cross and which may not. Freely permeable and totally permeable membranes would allow any substances to pass through, and an impermeable membrane is one through which nothing can pass.

19. **The correct answer is (D).** All cells can exist as distinct entities because of the cell membrane, which regulates the passage of materials into and out of the cell. Latex membrane does not allow the HIV virus to pass through the membrane, preventing HIV infection.

20. **The correct answer is (C).** The movement of materials against a concentration gradient requires energy. Consequently, this is called active transport.

21. **The correct answer is (D).** In active transport, the transport proteins move a solute against its concentration gradient. Active transport does not proceed spontaneously. It requires an energy (ATP) input.

22. **The correct answer is (B).** Although other mechanisms may be involved in passage of materials through a cell membrane, the passage of some materials, especially water, occurs by diffusion; *diffusion* occurs from regions of higher concentration of the substance to regions of lower concentration. Diffusion through a membrane, such as the cell membrane, is known as *osmosis*.

23. **The correct answer is (D).** Osmosis is the diffusion of water through a semipermeable membrane from a region of greater concentration to a region of lesser concentration.

24. **The correct answer is (B).** A cell is hypertonic when it has higher solute concentration and less water concentration compared to the outside solution, which has more water and less solute (hypotonic), so the water will diffuse inside the cell.

25. **The correct answer is (C).** Red blood cells have 0.9% salt solution whereas sea water has more solute than red blood cells, so it is a hypertonic solution compared to red blood cells.

26. **The correct answer is (C).** Osmosis is the movement of water across cell membranes from solutions of low solute concentration to solutions with higher concentrations of solute. The salt gives the water

a higher solute concentration than the frog's intracellular fluids. Therefore, the water concentration inside the cells would decrease as water flowed to the outside.

27. **The correct answer is (B).** The cell will lose water by osmosis down water's concentration gradient, which is lower in a hypertonic solution. Cells will swell when placed in a hypotonic solution. Osmosis cannot move water from an area of low concentration (the hypertonic solution) into the cell interior with its higher water concentration. Intracellular fluid is not hypertonic.

28. **The correct answer is (A).** A cell immersed in a solution with a lower concentration of dissolved materials (solutes) is in a hypotonic environment. The concentration of water is higher outside of the cell than inside. Under these conditions, water diffuses into the cell.

In-Depth Solutions for Acids, Bases, and Salts

1. **The correct answer is (A).** Hydrochloric acid has the weakest conjugate base (Cl) and therefore dissociates most thoroughly (virtually completely in water) and yields hydrogen ions in high concentration. The other compounds shown dissociate less completely because of their stronger conjugate bases.

2. **The correct answer is (D).** Acidity is determined by the concentration of hydrogen ions per unit of volume. Strong bases allow fewer hydrogen ions to escape in aqueous solution than weak bases, producing weaker acids (e.g., Cl⁻ is a weak base, making HCl and strong acid, whereas HCO_3^- is a strong base, making H_2CO_3 a weak acid).

3. **The correct answer is (B).** The conjugate base is obtained when the acid loses a hydrogen ion. Therefore, the conjugate base of carbonic acid, H_2CO_3, is the hydrogen carbonate ion (HCO_3^-).

4. **The correct answer is (B).** Acidic solutions contain hydrogen (H⁺) ions, while basic (or alkaline) solutions contain a basic ion, such as the hydroxyl ion (OH⁻).

5. **The correct answer is (D).** The pH of a solution is a measure of the negative logarithm of the hydrogen ion concentration (e.g., pH = –log [H⁺]). The H⁺ in question is 1×10^{-3}. Therefore, the pH = 3.

6. **The correct answer is (D).** Whether a watery solution is acidic or alkaline depends on its concentration of hydrogen ions (H⁺) in relation to hydroxyl ions (OH⁻). The ratio is expressed as the solution pH (the letters stand for potential of hydrogen). The pH scale ranges from O (most acidic) to 14 (most alkaline). Pure water is neutral with a pH of 7 because it has equal amounts of hydrogen and hydroxyl ions.

7. **The correct answer is (C).** Alkaline solutions always have a higher pH and more OH⁻ ions.

8. **The correct answer is (C).** Basic or alkaline solutions have fewer H⁺ than OH⁻ ions; their pH is above 7.

9. **The correct answer is (C).** Calcium oxide is the only compound in the four that is a metal oxide. Metal oxides form hydroxides (bases) when dissolved in water, but non-metal oxides form acids.

10. **The correct answer is (C).** Litmus turns red in acid solution. When phosphorus trioxide dissolves in water, phosphoric acid is formed. The other compounds named form alkaline solutions.

11. **The correct answer is (A).** Nonmetal oxides, such as oxides of sulfur, carbon, and nitrogen, can react with water to form acids. The equations below are examples:

$$CO_2 + H_2O \rightarrow H_2CO_3$$

$$SO_2 + H_2O \rightarrow H_2SO_3$$

12. **The correct answer is (C).** $NaHCO_3$ will combine with HCl (hydrochloric acid) to form NaCl and H_2CO_3. H_2CO_3 is a weaker acid than HCl; therefore, the net effect of replacing HCl molecules with H_2CO_3 molecules is to raise the pH of the solution. In causing this reaction, $NaHCO_3$ is acting as a *buffer*. NaCl is a salt that will have no effect on the pH of an HCl solution. The other two compounds are acids that would lower the pH of the solution further if added.

13. **The correct answer is (A).** A salt is an ionic compound that yields ions other than hydrogen ions (H^+) or hydroxide ions (OH^-) when it dissociates. Na_2CO_3 will yield neither when it dissociates into $2\ Na^+ + CO_3^{2-}$. $Ca(OH)_2$ will yield hydroxide ions when it dissociates, and H_2CO_3 will yield hydrogen ions. CH_3OH is a covalent compound and will not dissociate.

14. **The correct answer is (D).** The salt of a weak acid would hydrolyze in solution to remove hydrogen ions from water and leave an excess of hydroxide ions. $H_2O + A \rightarrow HA + OH$. The salt of a strong acid would hydrolyze by removing hydroxide ions from water. $NH_4^+ + H_2O \rightarrow NH_4OH + H^+$.

15. **The correct answer is (B).** Buffer systems resist significant changes in pH. When a strong acid or a strong base is added to a buffer system, they produce a weak acid and a salt or a weak base and water, respectively. In each case the pH would change only slightly.

In-Depth Solutions for Nuclear Chemistry

1. **The correct answer is (C).** Fusion is the combination of light nuclei to form heavier nuclei. Fused means combined in this context.

2. **The correct answer is (D).** Positron (with the mass of an electron) emission can occur, but protons are never expelled from the nucleus during radioactive decay.

3. **The correct answer is (C).** Nearly every unstable nucleus of an atom gives off one or more of the three kinds of radiation: alpha, beta, and gamma. *Alpha* and *beta* radiation are particle emission; *gamma rays* are shortwave energy rays similar to X rays, but stronger and more penetrating. Both gamma rays and X rays are more penetrating than alpha and beta radiation.

4. **The correct answer is (A).** The loss of an alpha particle reduces the number of positive charges by two and the atomic mass by four.

5. **The correct answer is (D).** Inasmuch as 100 milligrams will decay to 50.0 milligrams, the formula:

$$K = \frac{0.693}{t_{\frac{1}{2}}}$$

should be algebraically rearranged to solve for $t_{\frac{1}{2}}$ (half-life). To do this, multiple both sides by $t_{\frac{1}{2}}$ and divide both sides by K.

6. **The correct answer is (A).** The *half-life period* can be defined as the time needed for half of a given amount of a substance to undergo spontaneous decomposition. This varies with different radioactive elements. There would be 50 milligrams left of a 100 milligram sample of material with a half-life of eight days after the eight days pass.

7. **The correct answer is (B).** The half-life indicates that 100 years are required for of the sample to decay. Therefore, the 31.5 kg should be doubled 4 times.

8. **The correct answer is (C).** One gram of carbon-14 must go through four half lives to be reduced to 0.0625 grams. Four $\times$ 5.73$\times$10 years equal 22.92 $\times$ 10 years.

9. **The correct answer is (A).** The equation $E = mc^2$ demonstrates mass and energy are interchangeable and that small amounts of mass can yield large amounts of energy under specific conditions.

In-Depth Solutions for Organic Chemistry

1. **The correct answer is (D).** Organic molecules are molecules that contain carbon; they are found in living things.

2. **The correct answer is (B).** Propane contains three carbon atoms (n = 3) and eight hydrogen atoms (2n + 2 = 2(3) + 2 = 6 + 2 = 8).

3. **The correct answer is (A).** *Aromatic* compounds were originally named because of their odors; they occur in ring structure (molecular structure) and include such compounds as benzene, tolulene, and xylene.

4. **The correct answer is (D).** Single bonds between carbon atoms are longer and weaker than double bonds between carbon atoms.

5. **The correct answer is (C).** Triple bonds, sharing six electrons between two carbon atoms, provide each with a complete octet.

6. **The correct answer is (C).** A member of the acetylene series is unsaturated, with a triple bond between two carbons. The formula for acetylene, for example, is $H-C \equiv C-H$; because of the triple bond between two carbons, the number of hydrogens is reduced by two. Thus, the general formula is C_nH_{2n-2}.

7. **The correct answer is (D).** The ethyl alcohol of alcoholic beverages is produced by fermentation of monosaccharides; usually yeast is used to supply the enzymes needed, and the fermentation process supplies the energy needed by the yeast.

$$C_6H_{12}O_6 \rightarrow 2C_2H_5OH + 2CO_2$$

8. **The correct answer is (D).** Polarity is determined by differences in electronegativities between atoms involved in a bond. The difference in electronegativities between hydrogen and oxygen on a scale devised by Linus Pauling is 1.4, and between carbon and oxygen, it is 1.0. Carbon and hydrogen differ by only 0.4, and consequently, these atoms form nonpolar bonds.

9. **The correct answer is (B).** The functional group for ethers is R-O-R.

10. **The correct answer is (D).** Polarity is determined by differences in electronegativities between atoms involved in a bond. The difference in electronegativities between hydrogen and oxygen on a scale devised by Linus Pauling is 1.4, and between carbon and oxygen, it is 1.0. Carbon and hydrogen differ by only 0.4 and consequently form nonpolar bonds.

11. **The correct answer is (D).** The functional group contains carbonyl plus a hydrogen linked to the carbon (-CHO). The other connection of the carbonyl group is to another carbon atom. The groups -COOH, -OH, and -COOR are found in carboxylic acids, alcohols, and esters, respectively.

12. **The correct answer is (C).** A *ketone* is an organic molecule that contains a carbon atom double-bonded to an oxygen atom located between 2 other carbon atoms. The compound shown in choice (**A**) is an aldehyde because it contains a terminal -CHO group. The compound shown in choice (**B**) is an alcohol because it contains an -OH group bonded to a carbon. The compound shown in choice (**D**) is both an alcohol and an aldehyde.

13. **The correct answer is (D).** An *organic* acid is characterized by the presence of a carboxyl group
(—COOH), or $-C\overset{O}{\underset{OH}{\nearrow}}$. R represents a hydrocarbon group or a radical derived from a hydrocarbon.

14. **The correct answer is (A).** Note the R–COOH group, which makes the compound an acid, with the
R–NH_2 (amino group) in the alpha position.

In-Depth Solutions for Carbohydrates

1. **The correct answer is (D).** Glucose is a 6-carbon simple sugar and serves as a precursor of many complex compounds and as a building block for larger carbohydrates.

2. **The correct answer is (C).** The generic formula for carbohydrates is $(CH_2O)_n = 1C:2H:1O$ (1:2:1).

3. **The correct answer is (B).** Glucose and fructose are isomers because they are composed of identical numbers of carbon, hydrogen, and oxygen atoms, with different structural arrangements and different functional groups. The carbon-oxygen bonding in glucose and fructose form aldehydes and ketones, respectively.

4. **The correct answer is (A).** Dextrose is a *monosaccharide,* in that its molecule is composed of one "sugar unit" and cannot be hydrolyzed into simpler sugars. Lactose and sucrose are *disaccharides,* each yielding two monosaccharides by hydrolysis. Glycogen is a *polysaccharide* composed of a large number of monosaccharide units.

5. **The correct answer is (B).** Dissaccharides are double sugars and are formed by linking two simple sugars. For example, glucose and fructose form sucrose, a disaccharide (table sugar).

6. **The correct answer is (B).** Sucrose is a disaccharide that has nearly twice the molar mass of the other two sugars, which are monosaccharides.

7. **The correct answer is (D).** Carbohydrates consist of molecules made up of C, H, and O in a ratio of 1:2:1 (e.g., glucose is $C_6H_{12}O_6$). Polysaccharides are chains of three or more simple sugars (e.g., glucose).

8. **The correct answer is (C).** Liver glycogen, which is a polymer of glucose molecules, is called animal starch.

9. **The correct answer is (D).** Carbohydrates consist of molecules made up of C, H, and O in a ratio of 1:2:1, whereas cholesterol is a lipid (steroid) formed of four carbon rings.

10. **The correct answer is (D).** Wax is made of glycerol and fatty acids. It is nonpolar and insoluble in water. It is a lipid molecule.

11. **The correct answer is (B).** The enzyme sucrase adds a molecule of water (hydrolysis to the band that binds the monosaccharides forming sucrose) to decompose the dissaccharide to simple sugars.

12. **The correct answer is (B).** The divalent (Cu^{2+}) cupric ion is the species capable of receiving electrons.

In-Depth Solutions for Lipids

1. **The correct answer is (D).** Fats belong to the class of biomolecules known as lipids. These are high energy storage compounds.

2. **The correct answer is (B).** As a rule, polar liquids dissolve in polar liquids and nonpolar liquids dissolve in nonpolar liquids. Water is polar and oil is nonpolar. Therefore, oil and water are immiscible.

3. **The correct answer is (B).** Hard water contains calcium sulfate, calcium bicarbonate, or magnesium compounds. Soap is a mixture of sodium salts of organic acids that will react with these dissolved minerals. The insoluble calcium salt of the organic acid is precipitated out. This is the *scum*, or insoluble curdy material, that forms first. If enough soap is added, the soap acts as a cleansing agent after the first portion has precipitated the calcium ion.

4. **The correct answer is (D).** Oxidation is the loss of electrons or the loss of hydrogen from a substance. Therefore, the carbon-carbon double bonds of unsaturated fats are oxidized more easily than the simple bonds of saturated fats.

5. **The correct answer is (C).** Fats are glyceryl esters that, when hydrolyzed, yield glycerol and fatty acids. The cleavage of the ester linkage upon hydrolysis can be achieved by saponification, acids, superheated steam, or lipase, which is an enzyme that hydrolyzes fats. This may be represented as follows: $RCOOR' + H_2O \rightarrow RCOOH + R'OH$

6. **The correct answer is (D).** *Estrogen* is a complex of female sex hormones responsible for the appearance of secondary sex characteristics, such as widening of the pelvis, breasts, etc. Estrogen also functions in the menstrual cycle, preparing the uterus for implantation of the embryo.

7. **The correct answer is (D).** Aldosterone is the hormone released from the adrenal cortex that helps regulate sodium and potassium balance.

8. **The correct answer is (C).** Estrogen and progesterone promote development of female sex characteristics and regulate menstruation.

9. **The correct answer is (C).** Testosterone is the hormone that regulates male sex characteristics.

10. **The correct answer is (C).** Follicle stimulating hormone (FSH), a gonadotropin from the anterior pituitary, promotes follicle development and estrogen secretion in females, whereas in males it promotes sperm cell production. Testosterone is an androgen with most of its effects seen in males. Melanotropin activates melanocytes within the skin. Luteinizing hormone is responsible for initiating ovulation.

11. **The correct answer is (A).** Oxytocin promotes labor and delivery by inducing uterine smooth muscle contraction. Vasopressin (anti-diuretic hormone—ADH) promotes water reabsorption by the kidneys. Somatotropin (growth hormone) stimulates cell growth by accelerating protein synthesis. Prolactin participates in the stimulation of mammary gland development and promotes milk production.

12. **The correct answer is (B).** Cholesterol is an essential component of cell membranes. Microtubules are comprised of proteins arranged in very small diameter tubes. Ribosomes contain proteins and RNA. Cytosol is the fluid component of cytoplasm.

13. **The correct answer is (B).** Intensive research has demonstrated the basic structure of cell membranes is a mosaic of proteins in an outer and inner layer of phospholipids.

14. **The correct answer is (D).** Substance x is hydrophobic and nonpolar, so it can easily pass through the membrane due to the phospholipid structure of the membrane.

15. **The correct answer is (A).** Membranes are composed of a phospholipid bilayer of molecules interspersed with protein molecules. Any kind of nonpolar molecule, such as lipids or some other organic solvents (ether, acetone, etc.), can easily pass through the lipid layer of membrane.

16. **The correct answer is (B).** Small, uncharged lipid soluble molecules cross biological membranes more readily than other species. Water-soluble molecules cross if they are sufficiently small.

17. **The correct answer is (B).** Although stored and delivered by the gall bladder, bile is manufactured within the liver. The duodenum is the major site of digestion, where it receives digestive materials from the pancreas.

18. **The correct answer is (B).** The gallbladder stores bile and concentrates it up to tenfold. Bile is manufactured in the liver but not stored there. Neither the pancreas nor the ileum is involved in bile storage.

19. **The correct answer is (B).** Bile is released into the duodenum and breaks down the undigested fats into small droplets.

20. **The correct answer is (C).** The gallbladder stores and concentrates bile, which is used to break down fats into droplets and aids in the absorption of fatty acids and glycerol. Therefore, fats are restricted when there are diseases of the gallbladder.

21. **The correct answer is (C).** Bile, produced in the liver and released from the gall bladder, is involved in fat metabolism in the small intestine. Lipase is also involved in fat metabolism, but it is from the pancreas.

In-Depth Solutions for Proteins

1. **The correct answer is (B).** Protein, the body's vital building material, makes up the basic structure of all cells.

2. **The correct answer is (C).** Proteins are organic substances made up of several amino acids bound together.

3. **The correct answer is (B).** A *polymer* is a large, chain-like organic molecule formed by bonding together many smaller organic molecules of the same kind. A protein is a chain of many amino acid molecules joined to one another by peptide bonds. The amino acid molecules present, and the sequence in which they are joined, are considered the *primary structure* of the protein and determine its physical or chemical properties.

4. **The correct answer is (C).** Proteins are made up of units called amino acids. Amino acids join together by a covalent bond and form polypeptide chains (the primary structure of proteins).

5. **The correct answer is (A).** The basic building block of protein is the amino acid. Each amino acid is a small organic compound with an amino group, a carboxyl group (an acid), a hydrogen atom, and one or more atoms, called its R group. Total amino acids are 20. Examples of amino acids are tryptophan, alanine, glycine, aspartic acid, lysine, proline, etc.

6. **The correct answer is (B).** Two amino acid molecules can be bonded together by removing a hydrogen atom from the amino group ($-NH_2$) of one amino acid molecule and removing a hydroxyl group ($-OH$) from the carboxyl group ($-COOH$) of the other. The result is a molecule of water removed from the two amino acids and a covalent bond between the nitrogen atom in the amino group of one amino acid and the carbon atom in the carboxyl group of the other. This type of reaction is called *dehydration synthesis* and results in the production of water within cells that is called *metabolic water*. Creation of a bond between two amino acids can be illustrated as follows (R represents any of the 20 different side chains that can be included in amino acids):

Removing this water molecule will create a bond between the nitrogen and carbon.

7. **The correct answer is (A).** The bond formed between the carboxyl (RCOOH) and the amino group ($R-NH_2$) of two amino acids is a peptide bond, and the resulting compound is a depeptide.

8. **The correct answer is (B).** A *peptide bond* is the covalent bond that links the nitrogen atom on the end of one amino acid molecule with the carbon atom on the opposite end of another amino acid molecule. A peptide bond is formed by removing a molecule of water from the two amino acid molecules. Therefore, inserting a water molecule between two amino acids linked by a peptide bond will break the bond by putting a hydrogen atom back onto the nitrogen atom of one amino acid and a hydroxyl group ($-OH$) back onto the terminal carbon atom of the other amino acid. Breaking a bond between two molecules by inserting a water molecule between them is called *hydrolysis* and is a common way of digesting large organic molecules such as starches and proteins.

9. **The correct answer is (A).** Keratin is a protein that waterproofs and adds structural strength to skin. It is formed in the deepest layer of the epidermis and migrates to the surface with time. Melanin, carotene, and hemoglobin are pigments in the skin.

10. **The correct answer is (B).** Persons with sickle cell anemia carry a variant hemoglobin molecule in the red blood cells, instead of the normal hemoglobin A. The production of hemoglobin is under genetic control.

11. **The correct answer is (D).** *Sickle cell anemia* is caused by inheriting the defective form of a gene that codes for part of the protein portion of the hemoglobin molecule. The number of erythrocytes produced and the amount of hemoglobin in each erythrocyte are normal, but the altered structure of the hemoglobin molecules causes them to stick together in a manner that distorts erythrocytes into an abnormal crescent shape, making them less able to carry oxygen. Other types of anemia are due to insufficient hemoglobin within erythrocytes, too few erythrocytes, or erythrocytes that are abnormally small.

12. **The correct answer is (B).** The bulk of a hemoglobin molecule consists of 4 subunits, each one of which is a chain of amino acids. Iron is also present in hemoglobin but does not comprise as large a part of the entire molecule as the 4 chains of amino acids. Ferritin is an iron-containing molecule stored in the liver. Myosin is a major component of muscle cells.

13. **The correct answer is (D).** Anemia, a decrease in oxygen delivery, results from deficiencies in erythrocytes or hemoglobin within erythrocytes. Platelets, plasma proteins, and leukocytes do not participate in oxygen transport.

14. **The correct answer is (D).** Sickle cell anemia develops when hemoglobin forms long crystalline chains within erythrocytes, forcing the cells into their bizarre shapes. Pernicious anemia refers to abnormal destruction of red blood cells; iron-deficiency and aplastic anemias involve abnormal blood cell formation.

In-Depth Solutions for Enzymes

1. **The correct answer is (C).** Enzymes are called *organic catalysts*, in that they are produced by the organisms and catalyze reactions that occur within that organism. The other substances named do not function as catalysts.

2. **The correct answer is (D).** All enzymes are protein molecules, and protein is one of the organic molecules necessary for sustaining life.

3. **The correct answer is (B).** Enzymes are protein substances that act as biochemical catalysts. They affect the rate at which a specific reaction occurs.

4. **The correct answer is (A).** Enzymes are macromolecular proteins that serve as biological catalysts that lower activation energy requirements in cellular reactions.

5. **The correct answer is (D).** Enzymes are proteins that catalyze reactions of substances known as substrates.

6. **The correct answer is (C).** Each type of enzyme functions best within a certain temperature range. The chemical reaction rates decrease sharply when the temperature becomes too high. Humans usually die when their internal body temperature reaches 44°C (112°F) because it destroys the shape of the enzyme, and thereby, stops the metabolism.

7. **The correct answer is (B).** The slope (m) is 10.

$$\frac{\Delta y}{\Delta x} = \frac{75-35}{8-4} = \frac{40}{4} = 10$$

8. **The correct answer is (C).** In enzyme-catalyzed reactions, substrate molecules generally out number enzyme molecules, which allow the enzyme to become saturated and the reaction to proceed at maximum velocity.

9. **The correct answer is (B).** Amylase is an enzyme found in saliva, which begins the breakdown of starch. Lipase is involved in fat digestion; pepsin begins the digestion of proteins; and insulin is involved in glucose processes.

10. **The correct answer is (C).** Pepsin results in the breakdown of proteins into polypeptides. Lipase participates in fat digestion. Sucrase breaks sucrose down into monosaccharides. Amylase initiates the digestion of starch.

11. **The correct answer is (B).** *Trypsin*, an enzyme found in pancreatic juice, digests proteins, peptones, and proteoses to peptides. The digestion of peptides to amino acids also occurs in the small intestine under the control of the enzyme erepsin. Some protein digestion (to peptones and proteoses) occurs in the stomach, under the control of the enzyme pepsin. Ptyalin and maltase act on carbohydrates; steapsin, on fats.

12. **The correct answer is (D).** Carbonic anhydrase within erythrocytes reversibly catalyzes the reaction of carbon dioxide and water, a reaction which is very slow in the absence of the enzyme. Antigens, antibodies, and plasma enzymes do not influence the reaction.

In-Depth Solutions for Nucleic Acids

1. **The correct answer is (A).** Replication is the process of generating an additional molecule using a template molecule. It results in duplicating a molecule.

2. **The correct answer is (D).** Mutation is the change of a substance.

3. **The correct answer is (D).** Terminate means to end, while initiate means to begin.

4. **The correct answer is (C).** Immutable means unable to be changed. The antonym for immutable is changeable.

5. **The correct answer is (D).** Each nucleotide of a DNA molecule contains a molecule of deoxyribose sugar, a nitrogen base, and a phosphate group, which is attached to the 3' carbon of one sugar molecule and the 5' carbon of a second sugar.

6. **The correct answer is (B).** The small organic compounds called nucleotides have three components: a 5-carbon sugar, a phosphate group, and a nitrogen-containing base. Nucleotides are the basic building blocks of DNA and RNA.

7. **The correct answer is (A).** Genes are composed of segments of nucleotide bases of DNA molecules.

8. **The correct answer is (C).** The Human Genome Project, which is funded by the National Institutes of Health and the Department of Energy, was initiated in 1990 was completed in 2003.

9. **The correct answer is (D).** Data from the human genome project have shown the total number of genes in humans is close to 30,000.

10. **The correct answer is (B).** Genetic information is encoded in the particular order of nucleotides bases, which follow one another in DNA. RNA molecules function in the processes by which the genetic information in DNA is used to build proteins.

11. **The correct answer is (B).** In DNA and RNA molecules, adenine (A) must pair with thymine (T) or uracil (U). Guanine (G) always pairs with cytosine (C). This specificity is based on complementarity of molecular geometry, relative to the formation of double or triple bonds.

    mRNA = AGAUAU

    tRNA = UCUAUA

12. **The correct answer is (A).** Cellular proteins are synthesized at the ribosomes where codon triplets of m-RNA are paired with anticodon triplets of t-RNA to direct amino acid sequences.

13. **The correct answer is (C).** *Electrophoresis* is a technique that takes advantage of the fact that molecules in solution will migrate through a porous gel when an electrical current is passed through it. A mixture of differently sized molecules is placed into a depression at one end of the gel, and the current is turned on. Smaller molecules will travel farther through the gel than larger molecules, with the result that bands of molecules will accumulate at various distances from the starting point according to their size. These bands can be made visible by staining, allowing the results of one electrophoretic separation to be compared to another.

14. **The correct answer is (C).** Plasmids—small, circular self-duplicating DNA molecules isolated from bacteria—are major requirements for recombinant DNA technology.

15. **The correct answer is (A).** The restriction enzymes recognize and fragment the DNA (at the proper location) of the genes to be cloned.

16. **The correct answer is (B).** The splicing of DNA from two different sources is accomplished by the enzyme, DNA ligase, which catalyzes bonding by complementary base pairing.

17. **The correct answer is (D).** The general procedure for cloning includes: isolating DNA plasmid from a bacterium, isolating the DNA comprising the gene of interest from a different cell type, mixing the gene of interest with the plasmid to form a hybrid type of DNA called recombinant DNA, infecting a vector (bacteria or virus) with the recombinant DNA, and culturing the vector (containing the recombinant DNA) to produce many identical copies of itself (clones).

18. **The correct answer is (B).** The polymerase chain reaction (PCR) is an in vitro technique.

19. **The correct answer is (C).** DNA (deoxyribonucleic acid) is found in the nucleus of each cell and "stores" genetic information, or the genetic code. DNA replicates itself before mitosis or meiosis begins, and genetic information is distributed equally to the daughter nuclei. RNA (ribonucleic acid), transcribed on DNA, translates the genetic or hereditary information.

In-Depth Solutions for Nutrition and Digestion

1. **The correct answer is (A).** Caloric needs vary with age as well as with physiological state, activity, and the size of the individual. Both physical and mental growth and development are more rapid during infancy and early childhood than at any other time in one's life. Therefore, more food is needed in proportion to size to provide the needed calories of energy and to provide the materials for growth. A proper diet is essential for normal development.

2. **The correct answer is (A).** Dried milk is an inexpensive but valuable source of protein.

3. **The correct answer is (A).** Iron and calcium are the minerals most often deficient in the American diet.

4. **The correct answer is (A).** Milk contains only about 0.1 milligram of iron per cup. The recommended daily amount is 15–18 milligrams.

5. **The correct answer is (A).** Essential amino acids are those that cannot be manufactured by the body, and therefore, must be included in the daily diet.

6. **The correct answer is (A).** Digestion is the process whereby the enzymes in the body change food into simple substances that can be readily used by the body.

7. **The correct answer is (A).** Digestion in humans occurs primarily in the lumen of the stomach and small intestine where macromolecules are degraded to small molecules capable of being absorbed.

8. **The correct answer is (D).** The chemistry of life is closely dependent upon the chemistry of water. Water, in the presence of the proper enzymes, must be added to the bonds that bind monosaccharides in carbohydrates, amino acids in proteins, and fatty acids and glycerol in fats to degrade these polymers to their subunits.

9. **The correct answer is (C).** In humans, carbohydrate digestion begins in the mouth with the enzyme ptyalin, produced by the salivary glands.

10. **The correct answer is (A).** Fat yields the highest amount of energy per gram of any of the foods consumed.

11. **The correct answer is (A).** Energy production, in the degradation of foods, is associated with the release of hydrogen (oxidation). The ratio of hydrogen to oxygen is significantly greater in fats than in carbohydrates and proteins.

12. **The correct answer is (C).** Proteins are composed of chains of amino acids linked by peptide bonds, with primary, secondary, and tertiary structure.

13. **The correct answer is (C).** The amino nitrogen must be removed from amino acids in order to give them the basic hydrocarbon skeleton similar to carbohydrates. After the amino group is removed, the resulting compounds may enter the same metabolic pathways used by carbohydrates.

14. **The correct answer is (A).** Gastric and intestinal enzymes gradually break down the protein molecule into its separate amino acids.

15. **The correct answer is (C).** Many aquatic animals excrete ammonia as their nitrogenous waste, egg-laying animals excrete uric acid, and intra-uterine animals excrete urea.

16. **The correct answer is (C).** Nitrogenous byproducts of protein catabolism (including uric acid, creatinine, and urea) are not significantly reabsorbed and therefore appear in urine. Amino acids, vitamins, water, and electrolytes are all reabsorbed within the nephron.

17. **The correct answer is (D).** Urea is filtered from the blood by the kidneys and excreted in urine.

18. **The correct answer is (A).** Carbohydrates must be reduced to monosaccharides in order for intestinal absorption to occur. Disaccharides are enzymatically broken down in the small intestine, as are polysaccharides (glycogen and starch), which are broken down into oligosaccharides.

19. **The correct answer is (C).** Glucose is the end product of carbohydrate digestion.

20. **The correct answer is (C).** The enzyme helpers called coenzymes are complex organic molecules, many of which are derived from vitamins. These coenzymes can pick up hydrogen atoms that are liberated during glucose breakdown.

21. **The correct answer is (B).** A, D, E, and K are fat soluble vitamins. The other vitamins are water soluble, and calcium and iron are minerals.

22. **The correct answer is (B).** Water-soluble vitamins include vitamin C and vitamin B complex.

23. **The correct answer is (D).** Vitamin C is a water-soluble vitamin. Vitamins A, D, E and K are fat-soluble.

24. **The correct answer is (B).** Thiamine is another name for vitamin B_1.

25. **The correct answer is (D).** Vitamin D is referred to as the "sunshine" vitamin because it is formed in the body by the action of the sunshine on the cholesterol products in the skin.

26. **The correct answer is (D).** Scurvy is a disease caused by a deficiency of vitamin C.

27. **The correct answer is (A).** Vitamin K helps the liver to produce substances necessary for the clotting of blood.

28. **The correct answer is (D).** Prothrombin is produced in insufficient quantities if there is a deficiency of vitamin K; prothrombin produces thrombin, which acts as an enzyme to convert fibrinogen to fibrin, the mesh that traps blood cells, forming the clot. Thus, if vitamin K is deficient, prothrombin is deficient, and clotting is delayed or does not occur.

29. **The correct answer is (D).** Strict vegans, who eat no eggs or milk, are particularly at risk for vitamin B_{12} deficiency, since animal foods are the only source of this vitamin.

30. **The correct answer is (A).** A deficiency of vitamin B_{12} leads to *pernicious anemia*, a condition in which red blood cells fail to mature. Vitamin B_{12} is essential for the formation of red blood cells.

31. **The correct answer is (C).** The human body is composed primarily of oxygen, carbon, hydrogen, and nitrogen, which comprise approximately 96 percent of the body's mass, with about 15 other elements comprising the rest.

32. **The correct answer is (D).** The approximate percentages of oxygen, carbon, nitrogen, and potassium in the human body are: 65, 18.5, 3.5, and 0.5, respectively.

33. **The correct answer is (B).** Iron is a part of the hemoglobin molecule, the red, oxygen-carrying pigment of red blood cells. If iron is deficient, hemoglobin cannot be produced, and the formation of red blood cells is inhibited.

34. **The correct answer is (D).** Iron is a part of the hemoglobin molecule, and, as such, is essential for its formation. Hemoglobin is the oxygen-carrying pigment found in red corpuscles.

35. **The correct answer is (B).** Iodine makes up about 65 percent of thyroxine, a hormone secreted by the thyroid gland.

36. **The correct answer is (B).** Processed and canned foods are generally very high in sodium. Fresh fruits and vegetables (apples, spinach) are usually low-sodium, as are whole-grain breads.

37. **The correct answer is (A).** Biotin is also known as vitamin H.

38. **The correct answer is (E).** Satiated means feeling full and satisfied after eating. The antonym of satiated is hungry.

In-Depth Solutions for Energy Storage and Usage

1. **The correct answer is (D).** Photosynthesis is the process by which certain living plant cells combine carbon dioxide and water, in the presence of chlorophyll and light energy, to form carbohydrates and release oxygen as a waste product.

2. **The correct answer is (D).** *Photosynthesis* is the process characteristic of green plants by which carbon dioxide and water react to produce glucose and oxygen. The carbon dioxide used in the process enters the tissue from the air, passing in through stomata of the epidermis; oxygen produced by the reaction passes through the stomata into the air. This process occurs only in the presence of light, and therefore, continuous carbon dioxide removal occurs only in the presence of light.

3. **The correct answer is (B).** Carbon dioxide and water, in the presence of light and the chlorophylls, yield glucose and oxygen. The water is absorbed by the root system of green plants and translocated through xylem to the leaves and other parts, which are the photosynthetic organs. Carbon dioxide enters the plants through epidermal stomates.

4. **The correct answer is (A).** Water dissociates into H_2 and O_2. The O_2 comes entirely from the water molecules and the resulting H_2 is then used to reduce CO_2 to simple sugar.

5. **The correct answer is (D).** In the light phase of photosynthesis, H_2O is dissociated into $H_2 + O_2$. The O_2 comes entirely from the water molecules. The resulting H_2 is then used to reduce CO_2 to simple sugar.

6. **The correct answer is (B).** Photosynthesis is an endergonic reaction which uses $CO_2 + H_2O$ to produce simple sugar and O_2. It is driven by radiant energy from sunlight.

7. **The correct answer is (B).** During non-cyclic light reactions, energy from the sun drives the formation of ATP (which carries energy) and NADPH (which carries hydrogen and electrons). Oxygen is a by-product of photosynthesis.

8. **The correct answer is (A).** Oxygen is not required for photosynthesis to occur; however, green plants (with chlorophyll) need carbon dioxide, water, and sunlight for photosynthesis.

9. **The correct answer is (C).** Photosynthesis is the process by which an organic, energy-containing compound (sugar or glucose) is produced from inorganic materials. This is a process basic to the production of organic compounds that can be used for food. Algae and other green plants exhibit photosynthesis.

10. **The correct answer is (D).** Carbon dioxide (CO_2) comprises only approximately 0.04 percent of the earth's atmospheric gases. When this percentage increases, it prevents sun rays that strike the earth from radiating back into space. This produces warming of the atmosphere and the earth's surface, causing the greenhouse effect.

11. **The correct answer is (D).** *Combustion*, or burning, is a rapid reaction giving off much heat and light in a short period of time. The energy is expended more rapidly; thus, the reaction occurs at a higher temperature since the temperature of the combustible material must be raised to a combustion point. Cellular respiration occurs more slowly, at lower temperatures, and is controlled by enzymes. If cellular respiration occurred at combustion temperatures, cells would be destroyed.

12. **The correct answer is (C).** Cellular respiration oxidizes digested food molecules and produces CO_2 and H_2O as end products. It is a downhill reaction and occurs in both plant and animal cells.

13. **The correct answer is (D).** Reaction I is an uphill process that takes in radiant energy from the sun and converts it to chemical energy to combine CO_2 and H_2O to produce $C_6H_{12}O_6$ and O_2—which have additional energy.

14. **The correct answer is (C).** The reaction in II is downhill, with energy being released from $C_6H_{12}O_6$, generating CO_2 and H_2O with less energy.

15. **The correct answer is (A).** When one molecule of glucose undergoes aerobic respiration 36 ATP are produced. Anaerobic respiration (fermentation) yields 2 ATP per glucose molecule. Phosphorylation yields 10 ATP per molecule of glucose.

16. **The correct answer is (C).** Aerobic cellular respiration is more important to sustaining life because it produces 36 ATP compared to anaerobic, which only releases 2 ATP.

17. **The correct answer is (C).** This is the process by which a glucose molecule is changed to two molecules of pyruvic acid with the liberation of a small amount of energy. In the absence of oxygen, pyruvic acid can be converted to ethanol or to one of several organic acids, of which lactic is the most common. This process is called fermentation.

18. **The correct answer is (A).** In the absence of oxygen, plants and microbes convert pyruvic acid into alcohol and carbon dioxide by a process called fermentation.

19. **The correct answer is (D).** In the first stage, aerobic respiration (glycolysis) produces 2 ATP. In the second stage (Krebs cycle), it produces 2 ATP. During the final stage (electron transport chain), it produces 32 ATP.

20. **The correct answer is (A).** A molecule of glucose uses two molecules of ATP to supply activation energy to initiate the glycolytic pathway.

21. **The correct answer is (B).** Eukaryotic cells have high efficiency, relative to mechanical device. They are capable of extracting approximately 50 percent of the energy in glucose molecules for biological work.

22. **The correct answer is (D).** Muscles at rest transfer high energy phosphate groups from ATP to creatine, forming creatine phosphate and ADP. When needed, that energy is transferred back to ADP, resulting in production of ATP.

23. **The correct answer is (A).** The limited amount of ATP stored in muscle tissue supplies immediate energy for contraction when the ATP is converted to ADP. When the ATP is used up, it is recreated by energy from a reserve, creatine phosphate. When the creatine phosphate is consumed, oxidation of glucose to CO_2 and water provides energy for muscle contraction and for the resynthesis of creatine phosphate. Lactic acid can form anaerobically during severe muscle exertion, and its accumulation is partly responsible for the feeling of fatigue. Lactic acid diffuses out of the muscle tissue into the bloodstream and thus to the liver, where some of it is oxidized to produce further energy, and some can be converted to glycogen for carbohydrate storage.

24. **The correct answer is (C).** Glycolysis is the first stage of cellular respiration in which glucose is broken down step by step to form 2 pyruvic acid inside the cytoplasm.

25. **The correct answer is (C).** In the first stage of aerobic oxidation, glucose is oxidized to pyruvic acid; this is known as glycolysis. The two pyruvates resulting from glycolysis of a molecule of glucose enter into the Krebs cycle (Citric Acid cycle) and are oxidized to carbon dioxide and water.

26. **The correct answer is (A).** Carbon dioxide and water are the end products of the Krebs cycle (including the electron transport chain), which is the second phase of aerobic cellular oxidation of glucose for the release of energy.

27. **The correct answer is (B).** Each turn of the Krebs cycle generates 2 moles of CO_2 and H^+.

28. **The correct answer is (C).** As oxygen accepts electrons from the ETS, it acquires a -2 valence, which attracts $2H^+$ to form HOH.

29. **The correct answer is (B).** In cellular metabolism, glycolysis is an anaerobic process (requires no O_2), produces reduced nicotinamide adeninedinucleotide ($NADH_2$) and small amounts of ATP, and occurs in all types of cells.

30. **The correct answer is (C).** Cellular respiration, the utilization of oxygen and glucose, occurs within mitochondria. Rough endoplasmic reticulum is involved in protein synthesis and smooth endoplasmic reticulum is involved in synthesis of lipid materials. Golgi apparatus are involved in cellular packaging and delivery.

31. **The correct answer is (C).** Mitochondrion is the powerhouse of the cell. It produces maximum amounts of chemical energy (ATP) for sustaining life.

32. **The correct answer is (D).** Electron transport systems and neighboring channel proteins are the machinery embedded in the inner membrane that divides mitochondrion into two compartments.

33. **The correct answer is (D).** The first stage of respiration, the breaking down of the glucose molecules, always takes place inside the cytoplasm because all the enzymes are there for the reaction.

34. **The correct answer is (D).** *Glycogenesis* is the formation of glycogen from glucose; glycogen is the form in which carbohydrates are stored in animals. Glycogen is stored in larger quantities and more permanently in the liver; some is produced and stored temporarily in muscle.

35. **The correct answer is (D).** In active transport, energy (ATP) is needed. Mitochondrion is the main organelle for ATP synthesis.

In-Depth Solutions for Body Chemistry

Urine

1. **The correct answer is (B).** When the water intake is excessive, the kidneys excrete generous amounts of urine; if the water intake is lost, they produce less urine. The process is regulated by hormones.

2. **The correct answer is (B).** When water intake is excessive, the kidneys excrete generous amounts of urine; if water intake is lost, they produce less urine. The process is regulated by hormones.

3. **The correct answer is (A).** Cell membranes maintain their high external Na^+ and high internal K^+ concentrations by using ATP to energize membrane carrier proteins that form the sodium-potassium pump. This is an active transport mechanism, which moves both ions against concentration gradients.

4. **The correct answer is (D).** Aldosterone promotes the reabsorption of sodium within the kidneys. Thyroxine is a thyroid hormone and is not involved with sodium levels. Insulin and glucagon are involved in the control of blood glucose levels.

5. **The correct answer is (A).** Sodium is the major positive-charged ion within extracellular fluid and is a primary controller of water distribution. Potassium and magnesium are more concentrated within intracellular fluid. Calcium does not play a major role in water distribution.

6. **The correct answer is (D).** Following action potential activity, sodium and potassium are returned to their respective starting compartments by the active transport of the sodium-potassium pump. Since these movements are against concentration gradients, diffusion would be ineffective. Filtration is a process of capillaries, and osmosis refers to the movement of water.

7. **The correct answer is (A).** Exercise will cause perspiration, which evaporates from the skin. *Perspiration* is a liquid composed of water, salt, and a small amount of urea and is drawn from the bloodstream (from capillaries in the sweat glands) and released to the body surface through pores. Thus, rigorous exercise will cause the body to lose sodium (salt) and water.

8. **The correct answer is (B).** Oxytocin and antidiuretic hormone (ADH) are both products of the posterior pituitary. Oxytocin acts on uterine smooth muscle and mammary tissue, whereas ADH acts within the kidneys to promote water reabsorption. ADH is also a vasoconstriction, and these two actions combine to elevate blood pressure.

9. **The correct answer is (A).** Water plus creatinine and urea, which are nitrogenous wastes, are normal substances in urine.

10. **The correct answer is (D).** Normal urine has a low specific gravity.

11. **The correct answer is (C).** Protein substance in the urine is called albuminuria. It results from the failure of the kidneys to filter.

12. **The correct answer is (A).** Approximately 50cc of fluid per hour are filtered through the adult kidney. This amount varies with many factors such as perspiration, fluid intake, cardiac, and renal status of the patient.

13. **The correct answer is (D).** A symptom of diabetes is sugar (glucose) in the urine. *Benedict's solution* is a reagent used to test for the presence of reducing sugars such as glucose.

Respiration

1. **The correct answer is (C).** Gases diffuse down a partial pressure gradient, not a concentration gradient. Osmotic pressures and gradients refer to properties of fluids. Within normal limits, temperature plays a minimal role in gas movement.

2. **The correct answer is (B).** The amount of dissolved oxygen in blood is minimal compared to the amount carried as oxyhemoglobin. Carbon dioxide and bicarbonate are not physiologic oxygen transporters.

3. **The correct answer is (D).** Declining blood pH weakens the hemoglobin-oxygen bond (the Bohr Effect) and promotes oxygen release to tissues. Conversely, elevated pH will strengthen the bond. Changes in blood pressure do not appreciably affect oxygen unloading.

4. **The correct answer is (B).** Some CO_2 is carried in a dissolved form obeying Henry's Law, and some is carried by hemoglobin. Most CO_2, however, is carried as bicarbonate.

5. **The correct answer is (A).** Most of the carbon dioxide is transported in the blood plasma in the form of bicarbonate ion (HCO_3^-). *Erythrocytes*, *leucocytes*, and *platelets* (thrombocytes) are blood cells suspended in the plasma.

6. **The correct answer is (C).** A significant portion of carbon dioxide is transported within erythrocytes as carbaminohemoglobin. Blood gases are not transported by plasma proteins or leukocytes, nor are they transported on the surface of red blood cells.

7. **The correct answer is (B).** The diffusion of gas occurs across the thin, squamous epithelium lining of the alveoli.

8. **The correct answer is (C).** Carbon monoxide has a greater affinity than oxygen for hemoglobin. This binding blocks hemoglobin from binding oxygen for transport to body cells for metabolic needs.

9. **The correct answer is (B).** Hyperventilation can lead to a decreased arterial CO_2 level, which may result in alkalosis of respiratory origin. Extreme alkalosis may lead to convulsions and may produce seizure activity in epileptic patients.

Blood

1. **The correct answer is (B).** Arterial blood pH is critically maintained between 7.35 and 7. 45. Either a decline of pH (acidosis) to a level of 7.0 or an increase in pH (alkalosis) to 8.0 is potentially fatal.

2. **The correct answer is (A).** Although not as powerful as pH regulating mechanisms in the respiratory or renal systems, buffer systems essentially act at the rate of chemical reactions. The liver is not involved in the normal control of blood pH.

3. **The correct answer is (C).** Although generally slower, the kidneys are the most powerful and complete correctors of pH imbalance. Buffers are the most rapid but the weakest mechanisms. Although more powerful than buffers, respiratory mechanisms rarely complete a correction of pH imbalance. The bladder is not involved in acid-base balance.

4. **The correct answer is (C).** Plasma is the liquid portion of the blood in which corpuscles are suspended.

5. **The correct answer is (A).** *Plasma* is the liquid portion of the blood. *Whole blood* consists of plasma with its dissolved materials and the blood cells.

6. **The correct answer is (C).** Plasma is approximately 90 percent water, with the rest being comprised of numerous solutes, including proteins, nutrients, gases, hormones, ions, and products of cell activity.

7. **The correct answer is (B).** Aldosterone is released by the adrenal cortex in response to decreased blood volume, decreased blood sodium ions, or increased potassium ions.

8. **The correct answer is (C).** Blood pressure is the force of the blood exerted against the wall of the blood vessel.

9. **The correct answer is (C).** Insulin increases the permeability of the cell membrane to glucose, thus increasing the rate of glucose uptake by the cells from the bloodstream. Adrenalin promotes an increase in cardiac activity, respiratory rate, and the breakdown of glycogen (stored in the liver) to glucose, thus raising the glucose level of the blood. With physical activity, glucose is more rapidly utilized by cells, which must remove glucose from the blood, thus reducing the blood level of glucose.

10. **The correct answer is (B).** Glucagon elevates blood glucose by (1) promoting the breakdown of glycogen to glucose (glycogenolysis), (2) promoting glucose synthesis (gluconeogenesis), and (3) promoting the release of glucose from liver.

11. **The correct answer is (B).** Insulin is the hormone that lowers blood glucose levels through a number of mechanisms including activating receptors involved in the facilitated diffusion of glucose into muscle. Glucagon is the pancreatic hormone responsible for elevating blood glucose levels; epinephrine also elevates blood glucose. Renin, a hormone from the kidneys, initiates the renin-angiotensin system, which is involved in blood pressure control.

12. **The correct answer is (C).** The islets of Langerhans are located in the pancreas and produce insulin.

13. **The correct answer is (B).** Insulin is released by beta cells when blood glucose exceeds normal levels (70–110 mg/dL). Low blood glucose triggers the release of glucagon from the pancreas. Prolactin is not involved in the control of blood glucose. Elevated growth hormone levels cause elevations of blood glucose.

14. **The correct answer is (B).** Insulin increases the permeability of the cell membrane to glucose, thus enhancing the uptake of glucose from the blood by the cells. If insulin is deficient, glucose is not removed from the blood and utilized by the cells, resulting in an excess of glucose in the blood and leading to other symptoms of diabetes.

15. **The correct answer is (D).** Type I diabetes mellitus results from inadequate insulin production by the beta cells of the pancreas; high insulin levels would reflect just the opposite. Ineffective insulin receptors are associated with Type II diabetes. Both Type I and Type II diabetic patients display high blood glucose levels.

16. **The correct answer is (A).** Hypoglycemia results from too much circulating insulin and not enough glucose. Visual disturbances and decreased urinary output can also be indicative of this problem.

17. **The correct answer is (B).** Older, overweight patients are typically Type II diabetics, which tends to be familial, whereas Type I diabetes does not.

18. **The correct answer is (B).** Polyuria is one of the "three Ps" of diabetes: polyuria, polyphagia, and polydipsia.

19. **The correct answer is (A).** The symptoms of ketoacidosis are Kussmaul respirations (which are rapid and deep), nausea, and vomiting, as well as hot, dry, flushed skin. Hypoglycemia initially presents

with sweating, palpitations, anxiety, and tremulousness. Neuropathy, which usually affects the legs and feet in diabetics and involves sensory changes, as well as retinopathy, an eye condition that frequently affects diabetics, are not acute conditions.

20. **The correct answer is (C).** Lack of insulin leads to the utilization of fats for energy. Byproducts include ketone bodies, which are acidic and responsible for the acidosis seen in these patients. Diabetic patients are not alkalotic, and diabetes does not result in acid production through actions on the respiratory system.

Health and Medicine

1. **The correct answer is (B).** Histamine is an amine that is released when cells are injured.

2. **The correct answer is (D).** Caffeine is a stimulant found in coffee.

3. **The correct answer is (A).** Melanin is a dark pigment found in the cells of the basal layers of the skin. Skin color varies with the size and density of the melanin particles: the more melanin present, the darker the skin.

4. **The correct answer is (B).** The thyroid gland needs iodine for the formation of thyroxine.

5. **The correct answer is (A).** Radiation can cause *mutations,* or genetic change. Any change in the genetic code would alter or destroy the trait associated with that particular gene. This could involve changes in enzymes produced and other factors associated with cellular metabolism.

6. **The correct answer is (A).** Chemotherapy destroys new growth of cells in hair follicles, mucus membranes of the GI tract, and bone marrow, while killing cancer cells.

7. **The correct answer is (B).** The inability to pass Cl^- to the outside of cells in cystic fibrosis patients causes water to enter the cells by osmosis, making the intercellular mucus thicker than normal.

8. **The correct answer is (B).** *Phenylketonuria (PKU)* is a genetic disorder characterized by the inability of the affected person to convert excess molecules of the amino acid phenylalanine to molecules of the amino acid tyrosine. It is caused by inheriting a defect in the gene for the enzyme that catalyzes the conversion.

9. **The correct answer is (B).** Myoglobin is a red pigment that reversibly binds oxygen in muscle. It is structurally related to hemoglobin, the oxygen-carrying pigment found in blood. ATP and ADP are involved in phosphate energy group utilization.

In-Depth Solutions for Reading Comprehension of Science Topics

DNA

1. **(C)** DNA is as unique to individuals as a fingerprint because it determines the traits of an individual. This is the best conclusion from the reading passage.

   Answer A is insufficient because Johan Friedrich Miescher discovered DNA, Watson and Crick described the structure of DNA.

   Answer B is insufficient because the strands of DNA take the form of a double helix.

   Answer D is insufficient because nuclein is nucleic acid, the chemical family for DNA. Miescher did not analyze nuclein other than to identify that it was not protein.

2. **(C)** Enzymes are catalysts. This can be deduced because Miescher used enzymes to break down cells.

   Answer A, "enzymes are unstable," is not supported by information in the passage.

   Answer B, "enzymes are ineffective," is partially true because the enzyme could not break down the nucleus of the cell; however, the enzyme did break down cells, and therefore, was at least somewhat effective.

   Answer C, "enzymes are solutions," is not supported by information in the passage.

3. **(A)** Milk is a dairy product. The passage states that DNA controls "whether or not you have a tolerance for dairy products." Letter A is the best answer choice.

   Answers B, C, and D are insufficient because "DNA stores the blueprints for making proteins." Immunity to the common cold, individual's choice of residential location, and a person's inclination toward dishonesty are not determined by an ability or inability to make certain proteins.

4. **(B)** DNA is contained in the nucleus of the cell. The nucleus is the location Miescher originally discovered nuclein (nucleic acid).

   Answers A and D are incorrect. DNA contains nitrogen bases and sugar phosphate bands, not the reverse.

   Answer B is incorrect. DNA is not stored in the cell wall.

5. **(D)** Proteins are the building blocks of a cell because they can be broken down by enzymes and cells are also broken down by enzymes. It can be inferred that cells are made of proteins.

   Answers A and C are incorrect. The nucleus, not proteins, is the control center and the storage center of the cell.

   Answer B is incorrect. DNA, not proteins, is the blueprint of the cell.

6. **(D)** The molecular structure of DNA is a "spiral staircase."

   Answers A, B, and C are incorrect. No description is given for the structure of proteins, digestive enzymes, or RNA in this passage.

7. **(D)** Digestive enzymes are effective in breaking up protein.

   Answers A, B, and C are incorrect. Miescher discovered that digestive enzymes were ineffective in breaking up nuclein. Nuclein is nucleic acid. DNA is a nucleic acid; therefore it cannot be broken up digestive enzymes. DNA and nuclein are chemical compounds; therefore, digestive enzymes cannot break up all chemical compounds.

8. **(A)** The word "nucleus" refers to the round control center of the cell.

   Answers B, C, and D are incorrect. The walls of the cell are not mentioned in this passage. DNA is a strand of molecules with a helical shape.

9. **(C)** The scientific disciplines used in determining the structure of the DNA molecule include biology, chemistry, and physics. Watson Crick is a physicist.

   Answers A, B, and D were not mentioned in the passage.

10 **(B)** As described in this passage, model is a physical form representing a concept.

   Answers A, C, and D are incorrect. The model of DNA is not an exhibitor of fashion or a person on whom an artist bases his or her rendition. It is also not a miniature version of an existing object. DNA is small enough to fit inside the nucleus of a cell and would be difficult to make smaller in the form of a model.

Is It You're Fault If You're Fat?

1. **(D)** Lifestyle and genetic make-up are two causes of obesity.

   Answers A, B, and C are incorrect because they are too narrow and do not include the genetic component mentioned in paragraph B.

2. **(B)** "Genes may program some to feel hungry when they aren't and others to be less able to tell when they are full."

   Answers A, C, and D are not discussed in paragraph A or B.

3. **(C)** People who are more prone to insulin resistance are those who are likely to develop Type 2 diabetes.

   Answers A, B, and D are incorrect. Stomach problems are not mentioned. People who are overweight can be more prone to insulin resistance, not those with no weight problem. Insulin resistance does not lead to becoming obsese.

4. **(A)** People with gene defects must put forth great effort to overcome urges to eat more.

   Answers B, C, and D are incorrect. Paragraphs C and D do not mention overcoming diabetes or living a normal life. Being overweight comes easily to someone with a gene defect that causes them to overeat.

5. **(A)** The author's purpose in this passage is to inform.

   Answers B, C, and D are incorrect. The author is not using persuasive or entertaining language. The author is also not analyzing the problem.

6. **(B)** Researchers have found that the absence of a certain substance (leptin or MSH) in the body results in uncontrolled eating.

   Answers A, C, and D are incorrect. No mention is made of an overabundance of leptin or that overeating is a habit. MHS is a peptide, but that was not discovered by the researchers mentioned in these paragraphs.

7. **(B)** Obesity is a state of being significantly overweight.

   Answers A, C, and D are incorrect. Obesity is not fat. Obesity is not strictly a genetic condition. Obesity may be caused by factors other than overeating.

8. **(C)** The main idea of this passage is current research reveals a number of contributing factors to obesity.

   Answers A, B, and D are incorrect. Diabetes can be prevented with diet and exercise; however, overeating cannot be controlled by diet and exercise. Diabetes is related to obesity, but this is not the main idea of the passage. Obesity does not have a cure.

9. **(C)** From this passage, one can conclude that being fat may not necessarily be the fault of the obese individual.

   Answers A, B, and D are incorrect. Being overweight may lead to health problems, like diabetes, and children can have weight problems; however, these are not the best conclusions to make from this passage. Obesity can be attributed to many factors, including genetics.

Body Fluids

1. **(D)** Answer D is the best answer choice because it incorporates all of the other answer choices. Water is necessary for metabolism which involves dissolving and exchanging solutes as well as osmosis.

2. **(B)** The blood carries all nutrients to the cells and all waste products away from the cells. Blood is a transportation system.

   Answers A, C, and D are incorrect. Osmosis and diffusion happen in all types of cells, not just blood cells. Blood higher concentration of all solutes than other body fluids. Blood does not control metabolic processes within the cells.

3. **(C)** Concentration is a ratio of solute to solvent.

   Answers A, B, and D are incorrect. The placement of solutes and distribution of solutes may change the concentration value; however, those factors are not used to determine the concentration. The percentage of electrolytes to non-electrolytes does not change the concentration for an individual solute.

4. **(A)** The largest percentage of water in the body is for intracellularly (within the cell).

   Answers B, C, and D are incorrect. 15% of water goes to tissue spaces (interstitially) and 5% remains in the blood (intravascularly). Extracelluarly water would include water outside of the cells.

5. **(A)** In the process of osmosis, the primary factor is water because water flows from areas of low solute concentration to areas of high solute concentration.

   Answers B, C, and D are incorrect. The solute, ionization, and force are factors of lesser importance in osmosis.

6. **(B)** When placed in a hypertonic solution, a blood cell will shrink because the water will flow out of the blood cell and into the solution in an attempt to dilute the higher solute concentration in the solution.

   Answers A, C, and D are incorrect. If the blood cell were placed in a hypotonic solution, then it would swell and possible burst (destruction). If the blood cell were place in an isotonic solution, then it would not change.

7. **(B)** The primary factor in diffusion is solute because solute will pass from solutions of greater concentration into solutions of lesser concentration during diffusion.

   Answers A, C, and D are incorrect. Water is the primary factor in osmosis. Ionization and force are factors of lesser importance in diffusion.

8. **(C)** The solution with the highest potential for osmosis is the more concentrated (hypertonic) solution. Osmosis is a "sucking" process caused by a solution of higher concentration sucking water from a solution of lower concentration.

   Answers A, B, and D are incorrect. Isotonic solutions will not experience osmosis. Hypotonic solutions will lose water during osmosis and gain solute during diffusion because of the "force" of the hypertonic solution. If answers A and B are incorrect, answer D is also incorrect.

Blood Donors Available—No Thanks! I'll Do It Myself!

1. **(D)** The accelerated interest in autologous transfusion is due to a rise in fear of communicable diseases.

   Answers A, B, and C are incorrect. Blood donors may be in short supply, but this is not mentioned in these paragraphs. Blood donation does stimulate rapid production of new blood cells; however, this is not the reason that people are choosing autologous transfusions. This process will not necessarily save time in the event of emergency surgery.

2. **(A)** Saline is a cleansing or sterilizing agent because the blood is squeezed from the sponges into a container of saline before processing the blood to be returned to the body.

   Answers B, C, and D are incorrect. Storing one's own blood supply is a costly endeavor. Homologous transfusion may require blood typing and matching; however, that is not implied in this passage. Autologous transfusion is most feasible for elective surgery.

3. **(C)** Hemoglobin carries oxygen and carbon dioxide throughout the body. That is why doctors are researching ways to develop artificial hemoglobin for accident victims.

   Answers A, B, and D are incorrect. Autologous transfusions are not practical or easily accessible alternatives to blood transfusions in all cases. This passage has not indicated that all types of blood transfusion place a patient in a high-risk health situation. Lost blood is replaced quickly.

4. **(B)** A fact expressed in this passage is that suction devices and sponges are two surgical implements used in the collection of blood lost during surgery.

Answers A, C, and D are incorrect. Artificial hemoglobin is desireable for temporary transport of oxygen and carbon dioxide. Laser surgery is not discussed in this passage. Blood can be collected during surgery for autologous transfusions.

5. **(B)** Blood transfusions are guaranteed to be safe. Blood may be contaminated by disease, even if it has been analyzed by researchers.

Answers A, C, and D are incorrect. The article does not say that homologous transfusions are on the decline , autologous transfusions are always impractical, nor surgical patients who are transfused autonomously are assured a more rapid recovery.